Praise for *The Least of Th*

"Compassionate . . . a tribute to the many patients, living and dead, for whom Dr. Baxter has cared. It is also his gift to those of us who do not bear witness to the horrible scenes played out every day behind the doors of an inner-city AIDS ward; he gives us a chance to gain a little humility from listening to our dying brethren."
—*New York Times Book Review*

"The stories . . . possess a Tolstoyan power. . . . Clearly a man of immense sympathy and spiritual strength, Baxter here writes not only about the dying but also about the various ways that people confront death. . . . This book makes for powerful reading."
—*Washington Post Book World*

"This book does for the medical view of AIDS what Paul Monette's *Last Watch of the Night* did for the personal aspect."
—*Publishers Weekly* (starred review)

"Although Dr. Baxter's well-told stories concern AIDS patients dying in the dreariest of physical settings . . . they also show, with incredibly moving and honest compassion, a way to discover the human heart, basic values, and individual character. This book . . . inspires us to wake up and care for our suffering brothers and sisters."
—Thomas Moore, author of *Care of the Soul* and *the Reenchantment of Everyday Life*

"A powerful, challenging and highly spiritual book."
—*Catholic Register*

THE LEAST OF THESE MY BRETHREN

THE LEAST OF THESE MY BRETHREN

A DOCTOR'S STORY OF HOPE AND MIRACLES ON AN INNER-CITY AIDS WARD

Daniel Baxter, M. D.

A Harvest Book
Harcourt Brace & Company
San Diego New York London

This Harvest edition published by arrangement with
Harmony Books, a division of Crown Publishers, Inc.

Library of Congress Cataloging-in-Publication Data
Baxter, Daniel (Daniel J.)
The least of these my brethren: a doctor's story
of hope and miracles on an inner-city AIDS ward/Daniel J. Baxter.
p. cm.
Originally published: 1997.
"A Harvest book."
ISBN 0-15-600588-3
1. AIDS (Disease)—Patients—Hospital care—New York (State)—
New York. I. Title.
[RC607.A26B38 1998]
362.1'969792—dc21 98-4555

Design by Lenny Henderson
Text set in Bembo

Printed in the United States of America

First Harvest edition 1998
C E F D B

To Sister Pascal Conforti, director of Pastoral Care for the Spellman Center for HIV Related Diseases, an extraordinary friend and mentor, and a living saint.

CONTENTS

ACKNOWLEDGMENTS

THIS BOOK has many "authors"—namely, the many AIDS patients I have had the privilege to care for, in addition to the many families, friends, and colleagues whose influence and support have guided me in finishing this project. The pages that follow will indeed highlight many of these patients, but mention should be made here of the silent "supporting cast" behind my book.

First, my parents, Alfred and Eleanor Baxter, and my three siblings, Tim, Mary, and Faith, are responsible for my ability to feel—and *care* about—other people's suffering. Their love has largely made me the person I am today.

Presently physician in chief at Miriam Hospital in Rhode Island, Dr. Charles Carpenter was the chief of medicine and director of my internship and residency program at University Hospitals of Cleveland in the late 1970s. By force of his high professional standards, Dr. Carpenter is responsible for my "medical conscience" that has guided me over the past two decades. Even though he is never mentioned in the pages that follow, the Tall Man—as my residency colleagues and I aptly called Dr. Carpenter—has always cast a shadow of compassion over my AIDS work.

Over the past ten years, a series of chance meetings have spawned friendships that have profoundly affected my life. While I was working with the rural poor in the outpatient clinic at the Huntington, West Virginia, Veterans Hospital, Gary Workman helped me adjust to the emotional stresses of my "new life" in the Mountain State. Then, Dr. Vincent Mandarine's friendship propelled me to move on to Washington, D.C., where Gregory Harkness encouraged me to go to New York City. Greg's moral support—in addition to his advice regarding Catholic dogma—shines out from many of the following pages. My move to New York blessed me with two other special friends: Richard Wegener, whose spiritual values continue to inspire me, and Robert Weil, whose loving support for me and my AIDS work has always challenged me to take the moral "high road" in my daily personal and professional life. And supporting me during all of these moves have been my long-time friends, the Winchells and the Ginders.

Wendy Lipkind-Black, my agent, has believed unequivocally in my book from the start, and her kind words—and experienced business savvy—have guided me through the daunting terrain of the New York publishing world. And making possible this transition from manuscript to published book has ultimately been my editor, Leslie Meredith, who likewise saw beyond its "death and dying" and understood its real message of hope. Wendy Lipkind-Black and Leslie Meredith are what I like to call "real New Yorkers"—tough and street-smart, yes, but also with social consciences that genuinely care about the underdog, the disenfranchised. Special thanks must also go to Sherri Rifkin, Andrew Stuart, and Steve Boldt, who provided counsel and emotional support during the book's creation.

My many colleagues at St. Clare's deserve special mention for their friendship and invaluable sense of humor during our stressful work there, especially Anne Abbott, Father Jack, Dennis Ryan, Bruce Lockhart, Slavic Laczkowski, Cherilyn Thompson, and the extraordinary, dedicated nurses on Unit 3A. I am grateful for the support of my new friends at Casa Promesa and the Ryan Community Health center, especially Dr. Maribel Garcia-Soto and Barbra Minch, the best bosses I have ever had. I would also like to acknowledge Mr. Richard Yezzo for his devotion to the Spellman Center's mission.

Finally, my debt to both Sister Pascal Conforti and the many patients of the Spellman Center can never be calculated, let alone repaid. It is my hope that the pages that follow will clearly demonstrate the nature and extent of this debt.

PREFACE

"HOW CAN you stand to work with so much death around you all the time?"

The pleasant young woman's query was voiced sympathetically—even admiringly—and without recoiling disgust or disbelief. My questioner was a recently hired secretary in the Spellman Clinic, who had been randomly seated next to me at the Helen Hayes Dinner Dance, the annual black-tie gala thrown by St. Clare's Hospital that is a major fund-raiser for the financially strapped hospital. After the initial, introductory pleasantries, our conversation that evening—as often happens in such situations—quickly devolved into talk about our work at St. Clare's, especially the inpatient AIDS ward on which I was an attending physician.

Indeed, the topic raised by my dinner companion—AIDS deaths, primarily of impoverished inner-city patients—seemed starkly incongruous with our august surroundings: a glittering ballroom with innumerable balloons and glimmering tinsel hanging from the crystal chandeliers in midtown-Manhattan's Marriott Marquis Hotel, where hundreds of elegantly attired and coiffed patrons and staff of St. Clare's Hospital were seated at festively decorated tables for eight and entertained by a live band. The elaborate six-course meal and fancy place settings almost made the $350-per-person cost seem reasonable. Fortunately, my secretary friend, like most other nonphysician staff, was there on a discounted $50 ticket. Nonetheless, the New York glitz was light-years away from my dingy, desolate AIDS ward.

"I think what you do is *wonderful,*" my new admirer continued, again without affectation, "but I just don't think I could handle it. How do you do it day after day?"

Her question was difficult to answer succinctly without sounding like a Hallmark card, or dial-a-prayer sentiment. Her kind compliment about my "wonderful" work embarrassed me. I wanted to reply that I have always viewed my AIDS-ward work as a privilege, a job that always gave me more than I would ever be able to return—but in the past such protestations have sometimes sounded like false modesty. Being a doctor on an inner-city AIDS ward is a somewhat rarefied occupation, and I am often inclined to dismiss such flattery and inquiries with polite generalities, telling myself that people cannot really understand, unless they are right there with me, on my ward.

Yet, as on similar occasions, I wanted to answer my dinner acquaintance truthfully, and I felt frustrated at my inability to do so. I wished I could share with her the incredible insights and important lessons I experienced every day in my AIDS work—to give full dimension to my oft-repeated assertion that my job at the Spellman Center, St. Clare's AIDS service, was "the best job I've ever had" in my nearly twenty years of being a doctor. Moreover, on that gala evening, I wanted to reassure this young, attractive woman that death was not the fearsome enemy she seemed to imagine—that, if she harkened to the lessons of my AIDS patients, she could certainly "handle" the many AIDS deaths on a Spellman ward not only with equanimity but also, perhaps paradoxically, with *joy,* a transcendent *grace.* But the loud band and the surrounding table conversation that evening did not permit such an impassioned reply to her well-intentioned and profound question.

This book is intended, in part, to answer my dinner partner of that Helen Hayes benefit several years ago—to help her understand the extraordinary experiences I have had on my AIDS ward, in an attempt to illustrate the reassuring and life-affirming lessons that have so impressed me over my years there. Although enclosed herein are many patient stories of love, hope, and courage—stories that often transpire under the most desperate circumstances—this book is more than a collection of individual narratives. These patient stories—many of which are graphic and, at least on the surface, undeniably sad—serve a larger purpose of helping readers examine their own concerns about living and dying, since, as simple reflection tells us, *we are all ultimately HIV-positive* in this cumbersome experience called life.

The idea for this book blossomed gradually. Initially, I had planned to write only an article about an extraordinarily "impossible" patient I had cared for on my AIDS ward—the patient "Stephen Y." Stephen's case had so profoundly impressed—and emotionally exhausted—me that I felt compelled to share his extraordinary ordeal with others. Then, on a wintery Sunday afternoon, while listening to a sacred-music concert in a Midtown church—I was thinking about how I should approach the Stephen Y. story—the idea of writing a book about my AIDS ward flooded over me. Despite subsequent self-doubts, I have always persevered in my belief that this book's central message of hope *must* be told, both as tribute to my many memorable patients and as a cautionary message to others, such as my inquiring dinner companion at the Helen Hayes gala.

This book is based on my three and one-half years as a physician at St. Clare's Spellman Center for HIV Related Diseases, New York's largest Designated AIDS Center. The patients' names and circumstances have been changed, and to preserve confidentiality further, various details have sometimes been altered. Nonetheless, the pages that follow faithfully capture the essential nature of my experiences on my AIDS ward, that "crucible of despair *and* hope."

Verily I say unto you, Inasmuch as ye have done it unto one of the least of these my brethren, ye have done it unto me.

Matthew 25:40

1

IN MEMORIAM

Locus iste a Deo factus est inaestinabile sacramentum irreprehensibilis est.

This dwelling is God's handiwork: a mystery beyond all price that cannot be spoken against.

—TEXT FROM THE *MASS* FOR
THE DEDICATION OF A CHURCH

EVERY THREE months, with the change of seasons, a remarkable memorial service is held in the small chapel of St. Clare's Hospital, in New York City. Organized by the hospital's Spellman Center for HIV Related Diseases, this quiet service is intended, in the words of Sister Pascal Conforti, "to remember and celebrate the lives of our friends who have died and left us." The "friends" Sister Pascal speaks of were patients who have died from AIDS.

Centrally located on the ground floor of this aging hospital, the gemlike chapel of St. Clare's is neither opulent nor austere in decor. The founding Franciscan sisters designed it earlier this century as a place for both celebration and meditation, not religious ostentation. Inside the entrance doorway, competing for precious space, are a closet-sized confessional, a small basin of holy water, and a miniature Wurlitzer organ. Three tiny stained-glass windows filter amber light onto narrow wooden pews and reddish marble walls, which are crammed with small plaques denoting the fourteen stations of the cross. Framing a dramatic backdrop to the slightly elevated altar is a life-size oil painting of the glorification of St. Clare, St. Francis, and Our Lady of Mount Carmel. Despite the chapel's compactness—its total floor space might accommodate a regulation-size tennis court—an ethereal airiness suffuses throughout, perhaps because of the vaulted, blue ceiling, or because of the countless memories contained within its walls. Glowing warmly in late-afternoon sunlight, the chapel of St. Clare's is an inviting little space—a tranquil sanctuary from the drab and fearsome AIDS wards that surround it on all sides.

Participating in this quarterly memorial service is a cross-section of the Spellman Center staff: social workers, nurses, physicians, physician assistants, volunteers, secretaries, administrators. Although family and friends of the deceased are always invited, few are ever able to come, and most of the dozen or so people in attendance are Spellman employees. There are often more people at a Greenwich Village memorial service for one gay man than there are at St. Clare's thirty minutes of remembrance for the scores of its AIDS patients. As a Spellman physician, I go to these memorial services with both anticipation and trepidation: while exceedingly important to commemorate, the memories of former patients are sometimes too intense to revisit after an already difficult day on my inpatient AIDS ward. But I am always drawn to the Spellman memorial service because I feel it reminds me why I must be an AIDS doctor.

Starting at four with a quiet musical prelude on the chapel's organ, the service varies little from season to season. Sister Pascal gives a brief introductory welcome, followed by a staff member's reading of a meditative poem or short spiritual story. Then, to musical accompaniment, a procession of three or four staff marches to the altar, where they open the Book of Remembrance and light the memorial candle. Thereupon commences the heart of the service, the Reading of Names: quietly—*reverently*—the staff members alternate in reading aloud from the Book of Remembrance the names of the fifty or so patients who have died on the Spellman AIDS service over the prior three months. There are many names like Jose, Yolanda, Juan, Maria, and Hector—not the names of middle-class gay white males or of people listed in the *Times*'s obituary columns. These are society's poor and marginalized: the prisoners and ex-prisoners, the drug users and ex–drug users, the prostitutes, the homeless, the unwanted or the forgotten . . . unwanted or forgotten, that is, until they came to St. Clare's and its Spellman Center.

With the slow reading of the names of the dead, memories wash over the audience, engulfing everyone there in a spectrum of emotions. Each and every name evokes numerous, extraordinary remembrances and impressions: the patient's personality and idiosyncracies, or aspects of the patient's disease and course of illness, or his or her emotional response to dying from AIDS. No sooner does one name spark a cascade of memories than another name is called, and the process repeats. One name may recall the image of an ever-attentive mother—or sister or lover or spouse or friend—holding bedside vigil for his or her loved one. Another name might bring forth the haunting picture of an emaciated prisoner dying alone, far from his family in Puerto Rico. Yet another might evoke that patient's engaging smile and sense of humor, as she would playfully joke with her doctors and nurses. As the Reading of Names solemnly proceeds, the mind recalls random, even trivial images—a family photo on a patient's bedside stand, a patient's favorite dress or colorful turn of phrase, a handmade quilt that always covered a patient's bed, a small picture of the Blessed Virgin a worried mother taped on the wall of her only son's room, or one of a thousand other minor details of a patient's life on the Spellman service.

Some of the names are of patients who died on Unit 3A, my own AIDS ward; others are of people who passed away on adjacent Spellman floors. Even so, these latter patients are largely well known to me, since I may have cared for them on an earlier admission or heard their cases discussed at Spellman's TB rounds or morning report. Indeed, as Sister Pascal says, most Spellman patients have become friends to the staff.

"Hector R." . . . the crunching sound of ribs cracking during a cardiac resuscitation so violent that his teddy bear is inadvertently pushed off the bed onto the floor, into a pool of mal—

"Joey S." . . . taped on the walls around his bed are the many handmade get-well cards and drawings from his little nephews and nieces, whose hope for their uncle Joey's recovery contrasts so graphically with the bloody sce—

"Ricardo C." . . . his worried mother's refusals to understand her only son's terminal condition, as she would endlessly pace about his hospital room all hours of the day and night, nervously reacting to his every moan, his every—

"Rosa M." . . .

And so it goes: during each memorial service, my mind is brusquely whipsawed from one universe of memories to another, as the somber, rhythmic recitation of names proceeds with three-second regularity.

After the last name is read from the Book of Remembrance, an almost palpable emotional exhaustion ripples through the chapel. We sense that the awesome memories we just felt could only be encompassed by an infinite intelligence and love, far beyond mortal grasp. As the four staff close the

Book of Remembrance and file back into the audience, I am relieved that the emotional roller coaster has ended, but am uneasy over the truncated memories that I feel I *must* yet ponder and take to heart.

After the Reading of Names, someone else reads another poem or brief Bible passage, and then two or three staff members or relatives come forward and recount "Stories of Remembering"—anecdotes or impressions about patients who have most affected them. A Spellman volunteer relates how her acceptance and understanding of death were enhanced by a dying patient's courage. A social worker speaks of a previously fragmented family reunited by a dying patient's love. A physician assistant, an avid Yankees fan, tells how he single-handedly took another fan—his patient with terminal, AIDS-related lymphoma—from his hospital bed to a Yankees game a week before he died. A family member, barely holding back tears, relives happier times, when a loved one was active and well. Without exception, Spellman patients are remembered as vibrant individuals—never as diseased people dying from AIDS in hospital beds, but rather as *people who have much to teach about life and living.*

Finally, the program closes with a song or hymn, followed by a brief prayer by the hospital priest, Father Jack. The entire service lasts no more than thirty minutes or so. It is unrehearsed and unfinished; at times awkward gaffes occur. Sometimes the background music warbles off-key—the Wurlitzer is *very* old—or the 1950s public address system sputters on and off. Occasionally, a speaker's meditative story will ramble on, never quite coming to the point. During Stories of Remembering, tears frequently cut short a friend's reminiscences. Yet as simpleminded as it might seem to a cynical outsider, an unspeakable sadness and intensity permeates every Spellman memorial service.

Immediately after every service, a punch-and-cookies reception is held in the hospital's ancient cafeteria, where relatives and friends in attendance can visit with the staff who cared for their loved ones. The reunion can be bittersweet: most of the friends and family have not been back to St. Clare's, or spoken with their loved one's caregivers, since his or her death. There are often more tears at the reception than in the chapel, as memories of the deceased are relived. But always standing close by, ready with a supportive word—or a hug, if necessary—is Sister Pascal, the director of pastoral care and Spellman's guardian angel. Pascal radiates transcendent strength and serenity, renewing the flagging spirits of family and staff alike. As I watch her move among everyone at the reception, I marvel at her all-inclusive compassion. It is like seeing Schweitzer quietly working in his Central African hospital, or Jane Addams in Hull House. Pascal likes to boast that, for all its many problems, St. Clare's Spellman Center has "soul," due in no small part to her devotion to its patients and its healing mission—"a mystery beyond all price that cannot be spoken against."

After twenty minutes or so, the reception breaks up, and the family and friends return home, to continue with their lives, which sadly often entails looking after yet another friend or relative who has AIDS and may someday have his or her name called out from the Book of Remembrance. The staff drift back to their floors, to finish up the day's work.

I leave each memorial service with a feeling of incompleteness—an ill-defined sadness, an uneasy longing. There is *never* enough time to pause and relive each deceased patient's story, to apply each patient's experiences with AIDS to my own life. Walking home, I feel compelled to revisit many of the names recited earlier. Hours later, the Reading of Names still reverberates in my head as I reflect upon the protean lessons about life each name of the dead speaks to me.

2

A DAY IN AN AIDS DOCTOR'S LIFE

"Hector R." . . .
"Joey S." . . .
"Ricardo C." . . .

There are tumultuous days on my AIDS ward when the tranquillity of the quarterly memorial service, especially its Reading of Names, seems another galaxy—indeed, another universe—away. These difficult days can sometimes be so distracting and contentious that I must later pause and remind myself why I am an AIDS doctor, why I work at a crazy place such as Unit 3A . . . as on a certain hot Monday in late August . . .

IT'S ONLY 8 A.M., and as I walk onto the stifling AIDS Unit 3A to begin my rounds, all hell is about to break loose.

No sooner do I reach the unit's cramped nursing station than the head nurse pulls me aside to report that Jose T.—a new admission from the previous night—is confused, short of breath, and febrile to 103 degrees. For whatever reasons, the medical intern covering the unit last night ignored the night nurses' repeated calls for help, and Jose's worsening condition is now *my* problem to face first thing this morning.

Before going in to see Jose for the first time, I grab his still-thin hospital chart and quickly scan it to discover that he was paroled from an upstate prison less than a week ago. While an inmate, he had been treated in the dispensary for vaguely described complications of AIDS, including malnutrition and high fevers of unknown cause. The records from the prison are sketchy, but they indicate that the Department of Corrections—as it so often does with its paroled prisoners with AIDS—unceremoniously deposited Jose into an SRO (single-room-occupancy) hotel just last week, without medications or any arrangements for medical follow-up, and with only fragmentary medical records. Simply put, Jose was no longer the Department of Corrections' problem: he had just fallen through a hole in New York's increasingly frayed social service safety net.

Too weak to care for himself properly and becoming increasingly sicker, Jose had called 911 yesterday afternoon. Responding to his distress, an EMS squad picked him up during this intense summer heat wave and brought him to the emergency room of St. Clare's, the nearest hospital. Whether Jose has close family or, if so, whether they were even informed about his release from prison, let alone his hospitalization, is unclear from his chart—and doubtful. Before going into his room to see him, I scour the ER notes for what little information the staff there was able to obtain before he was shipped up to Unit 3A during the early evening hours: he is thirty-one years old, Hispanic, and an ex–intravenous drug user (which was undoubtedly how he acquired HIV). His chest X ray in the ER reveals a pneumonia, and his preliminary ER blood tests show evidence of malnutrition. Even before I lay eyes on him, I instinctively know that Jose is seriously ill from AIDS and has been abandoned, like almost all of my other sixteen AIDS patients on Unit 3A.

But Jose T. and his worrisome symptoms are not the only problems vying for my immediate attention so early this Monday morning on Unit 3A. My appearance on the unit has attracted the attention of the nurses as well as several patients and other claimants to my time. This is a rite of my morning rounds, as predictable as the appearance of daylight just a few hours before. Circling the nursing station is a gathering crowd of eager and familiar supplicants who begin to close in on me—like predators on prey, I think to myself as I see them approaching out of the corners of my eyes. It is way too early in the day to start feeling paranoid and overwhelmed, I tell myself, but I am already becoming claustrophobic at the tiny open nursing station, hemmed in by the approaching crowd of petitioning patients and inquiring family, all wanting an audience with the doctor. The unit is starting to feel uncomfortably close and muggy—there is no air-conditioning in most Spellman hallways—and my forehead starts to perspire slightly, both from the heat and from my exasperation at knowing that my intention of seeing Jose T. without being interrupted is doomed.

Moreover, I am upset with myself because of my annoyance with my patients: I *do* care about them and know that their myriad problems are monumental compared to mine. But Jose and his problems await; they are urgent and I am only one person, not the half dozen persons I feel I should be right now. I studiously try to ignore the four or five people congregating around me and intently focus on Jose T.'s chart, but to no avail—politeness is not the rule on Unit 3A, and the procession of patient requests, demands, and complaints begins.

To start, one of my patients, an active IV heroin user whom I admitted last week for hepatitis, sees me at the nursing station and is promptly in my face, loudly insisting on an increase in the dose of methadone I had started her on to ease her withdrawal from the heroin.

"You gotta increase it, Doc, or else I'll get some myself," she snarls, not reticent at all about reminding me how easy it is for Spellman patients to get illicit drugs in the hospital.

I continue to stare at Jose T.'s chart, not acknowledging her complaint; then another patient, Roberto Q., tugs on my white coat to complain indignantly about his breakfast: "I ordered eggs, and they sent me oatmeal!" he loudly gripes, as if I had the time—let alone the clout—to pick up the phone and rectify such all-too-commonplace menu errors, which especially seem to have vexed Roberto recently, almost to the point of obsession. I pretend not to hear Roberto's lament, and he walks away, muttering to himself.

Worried family members already visiting their loved ones on 3A early this morning are hovering close by the nursing station. Mrs. R., whose only son, Hector, is slowly dying from AIDS-related lymphoma, timidly gestures to me that she would very much like to talk with me about her son's condition. As during the past weeks, I can see in the lines of her face her desperate need for reassurance. Before I can signal back to Mrs. R. my intention to talk with her later, another nurse, who is handing out patients' breakfast trays, matter-of-factly reports that Joey S.'s rectal bleeding has been increasing since last night and that a dozen or so members of his tightly knit Italian-American family are keeping vigil in his room. This same nurse also mentions, again matter-of-factly, that Thomas A., one of my highly infectious TB patients, is dressed in street clothes and insisting on leaving the hospital this morning to pick up his welfare check in the Bronx—which would expose everyone around him to his contagious tuberculous slime. I grimly nod to the nurse that I have the messages and slam Jose's chart back into the chart rack, knowing that if I do not try for an escape to Jose's room right away, I will be trapped all morning at the nursing station.

The onslaught continues, with fresh reinforcements. There is Mr. Y., who is supposed to be discharged today and is impatiently pacing the hallway next to the nursing station, glaring at me because he expects me to drop everything else and take care of his discharge paperwork and drug prescriptions first thing. Almost on cue, a severely demented patient—actually a stray wandering onto 3A from an adjacent Spellman unit, dressed only in badly stained underwear—is halfway down the hallway and yells out to anyone within earshot that someone has stolen his cigarettes. "If I find out who did it, I'll beat the fuckin' shit out of 'em," he bellows, right before slipping on the corridor floor and almost falling on his ass. A nurse's aid is thankfully nearby and breaks his fall, patiently escorting him back to his unit for what is probably the twentieth time today alone.

The drama of *M*A*S*H,* which captured the fancy of millions of television viewers in the 1970s and 1980s, is mild compared to the circus that unfolds with my arrival on the floor.

Harold M., a patient with kidney failure who is scheduled to have important surgery later in the day, sees me at the nursing station on his way to the shower and nonchalantly announces that he has decided not to have the surgery after all—never mind the urgent necessity of the operation and the *herculean* efforts I went to last week to convince the reluctant surgeons to do it. A Spellman social worker in charge of discharge planning has just arrived on the unit, arms full with various medical forms, and is about to pepper me with several questions about the home health care needs of a patient who we hope will be discharged in a few days. And finally the head nurse returns and tells me that Ciano L., my patient with previously life-threatening fungal meningitis, has been responding so well to antibiotic therapy that he was strong and alert enough last night to wander off to an adjacent ward, where hospital security officers found him shooting up heroin in another patient's bathroom, with a supply of dirty needles and syringes.

This is a typical day on Unit 3A, my AIDS ward, that improbable crucible of despair and hope.

St. Clare's Hospital had always existed in the shadows of both Midtown's bright lights and the city's medical establishment. Founded earlier this century by a single-minded Mother Mary Alice and her six Franciscan sisters—all six of them her younger nieces, whose labors also provided food for the hospital from their order's farm in New Jersey—St. Clare's has had a checkered, occasionally desperate past, transforming and reinventing itself several times to serve the changing needs of Hell's Kitchen, that polyglot community immortalized by *West Side Story.* Hidden away in the middle of a block of ancient low-rise tenements, the hospital is nowadays surrounded by a neighborhood that has retained the frenetic energy of its immigrant past, despite its proximity to Midtown. Hell's Kitchen rarely attracts tourists to its cheap ethnic restaurants, its countless Korean groceries, and its Pakistani smoke shops. Although as safe as any New York neighborhood, the environs of St. Clare's have often seemed more hospitable to street drugs, panhandlers, and the homeless than the brigades of tourists from Kansas and the like.

Despite its relative obscurity, St. Clare's AIDS service was New York's first Designated AIDS Center and had, in fact, saved St. Clare's from extinction, enabling it to keep open its fledgling non-HIV medical and surgical floors. Occupying half of the hospital's two hundred beds and scattered throughout half a dozen floors, the Spellman Center had grown since its founding in 1985 to become New York's largest Designated AIDS Center, a distinction of more than semantic significance. A New York State Designated AIDS Center, by statute, had to provide comprehensive, multidisciplinary AIDS care—from social service support to pastoral care to substance abuse counseling to psychiatric services to volunteer programs to the requisite medical and nursing expertise in HIV. St. Clare's emergence from bankruptcy in 1982 coincided with the onset of the city's AIDS epidemic, which initially targeted New York's gay community, but by the mid-1980s, had made alarming inroads in IV drug users, the homeless, and minority women. All of these newer victims comprised a less cohesive, already underserved population that had few advocates and even less political clout. Whereas the besieged gay community had begun marshaling its resources into health education, social services, and emotional support for its AIDS casualties, the even more marginalized "second wave" of people with AIDS—appearing as a tidal flood by the late 1980s—had literally nothing, including any sense of community. Few hospitals, if any, were eager to serve such patients, especially when they brought with them the festering anger of years of abuse and neglect.

Having cared for several generations of Hell's Kitchen immigrants, St. Clare's was well positioned, by both its location and its Franciscan tradition, for this extraordinary population. Sister Mary Alice and her nieces would probably have marveled at St. Clare's transformation into a frontline battlefield in an epidemic of biblical proportions. The Franciscan sisters' quiet little hospital had been swept into the churning vortex of AIDS and urban poverty.

Located on the third floor of the hospital, Unit 3A is one of six AIDS floors that together comprise the inpatient arm of the Spellman Center for HIV Related Diseases. Although 3A's seventeen beds—nine of which are in private rooms—do not strictly qualify it as a "ward," the unit has the stark look of a charity hospital ward from an earlier age of American medicine. The overwhelming impression a visitor gets is that this tiny hospital unit was not originally built for its present function.

The cramped, compact nursing station, situated in the middle of the unit, was originally built with shelves too small to accommodate the countless hospital forms, bulging patient charts, and ever-changing hospital policy manuals that now dominate modern American medicine. The narrow hallways on both sides of this nursing station harken back to a simpler time, before medical technology cluttered the limited floor space with portable machines and carts. Many decades ago—before AIDS and the onset of "modern medicine"—3A was a quiet corner of St. Clare's where, with a minimum of technology and paperwork, dedicated nuns and doctors with the old-fashioned demeanor of Norman Rockwell subjects did their best to care for the immigrants of Hell's Kitchen. Nowadays 3A has become a cacophonous din of humanity, where everyone—patients, staff, visitors—dodges nursing medication carts, wheelchairs, X-ray and suction and EKG machines, portable oxygen tanks, laundry hampers, housekeeping carts, and maintenance tool carts, all the while jostling one another. This congestion is further compounded by 3A's central location in the hospital, which makes it a crossroads for hospital pedestrian traffic. There is no place for anyone—doctor, nurse, physician assistant—to escape for peace and quiet.

"State-of-the-art" is not ever used to describe 3A; rather, "crumbling infrastructure" or "bad case of the 1930s" feels more appropriate. The unit is in a constant state of repair, with maintenance workers fixing one broken item after another—adding to the tumult. Summers always feel more oppressive with the lack of hallway air-conditioning, but winter brings on its own problems, since the unit's aged thermostat regularly goes haywire, freezing the staff until repairs are finally made, often weeks later. The most modern aspect of the hallways is the fluorescent ceiling lights, but the underlying electrical wiring is ancient. Several summers ago, a fire in a fuse box forced relocation of the entire unit to another part of the hospital for a week until emergency repairs patched things up. That same summer also saw an adjacent Spellman floor closed by a foot-high mini-flood from the rupture of its antique plumbing. Such emergency evacuations of Spellman units to safer quarters are reminiscent of wartime MASH units, and these disasters, although not frequent occurrences, can force an entire inpatient unit to pack up everything right down to paper clips and Band-Aids and move to another vacant floor in only a few hours.

Less cataclysmic problems afflict 3A on a more regular basis: ceilings in patient bathrooms crumble and collapse, window frames in patient rooms disintegrate, toilets overflow, doorknobs fall off the doors, and the elevator breaks down almost daily. The resultant bedlam can be deafening—sounds of saws, hammers, and electric drills regularly form a crude obbligato over the unit's baseline commotion of patients clamoring for pain pills, nurses yelling at recalcitrant TB patients to get back into their isolation rooms, and doctors cursing at the lab's latest screwup. One visitor to Unit 3A once remarked that he felt he was in a third-world country's health infirmary, not a midtown-Manhattan hospital.

Contributing to this third-world impression are basic problems such as hygiene. A paucity of showers for the patients—3A has only two for seventeen patients—as well as a lack of public sinks for hand washing by the staff threatens observance of the time-honored axiom that simple hand washing controls hospital spread of infection. The only staff sink is in the staff bathroom located far from the patients' rooms, so doctors and nurses have only the patients' bathrooms for washing their hands—a somewhat unprofessional, albeit necessary, alternative. Even then, more often than not the patient's bathroom does not have soap or paper towels, since housekeeping is as overburdened as every other department.

Indeed, the problems with the housekeeping services go far beyond supplying patient bathrooms with soap and towels. The unit's pathetic lack of cleanliness was once underscored when I went into a patient's dirty room and saw his elderly mother on her knees, hand-scrubbing the filthy toilet bowl, which had not been cleaned for over a week. This was the same toilet that six months earlier had remained clogged with stool and urine for four days while the housekeeping and maintenance departments bickered over who was responsible for cleaning up the mess. As in any bureaucracy, the hospital's housekeeping service is overstaffed with supervisors. Whenever doctors or nurses scream about a particularly egregious lapse, a cadre of supervisors often arrives on the scene, clipboards in hand, to watch a lone grunt housekeeping worker do the job. Even the hospital administration can claim some credit for the problems with patient cleanliness. In a cost-saving step, it decreed that bedsheets were to be changed only three times a week, not daily as in most other hospitals—fortunately, most floor nurses ignored this directive.

The bleakest patient room on Unit 3A, room 314, has appropriately been nicknamed the Valley of Sorrows. This squalid room has four patients crammed into a space suitable for only two, and the sight of oxygen tanks, suction machines, bedside commodes, and countless other medical apparatus creates an impression of a tenement backyard, strewn with garbage and broken kitchen appliances. The torn and stained patient privacy curtains and the bags of IV fluids tenuously suspended from off-kilter IV poles evoke wash lines with flapping laundry hung across a slum-dweller's urban wasteland. The room has no overhead lights, and the small, often flickering fluorescent bulb above the head of each bed barely illuminates the top part of the bed, making nursing duties and physician examination of the patient almost impossible, since little sunlight ever gets through the small windows caked with years of grime. Despite such inconveniences, the most incredible aspect of this four-bedded room is that *it has no bathroom*—across the hallway is a three-by-four-foot closet with a tiny sink and toilet, and no shower. A bathroom on an Amtrak train is far more spacious. When one of the patients in room 314 has intractable, watery diarrhea, neither room 314 nor its "bathroom" is a pretty sight.

3A's most hopeless and most terminal patients somehow end up in room 314, because healthier, more alert patients often loudly object to being placed in such a pit. The end-stage, bedridden patients are too detached from the external world to complain about this affront to human dignity. Caring for patients under such conditions creates anger and despair, especially when so much money is spent on renovating the hospital administration's offices. Someday, perhaps a small fraction of the proceeds from the Helen Hayes Dinner Dance might be diverted for a proper bathroom for the Valley of Sorrows, or at least for overhead lights and clean windows.

Such were the conditions of the hospital AIDS ward that I worked on when Jose T. arrived that hot August weekend.

Despite the impending deaths of Joey S. from rectal bleeding and Hector R. from terminal cancer, I decide that Jose T. should still be my first priority. Before seeing him, however, I quickly dispense with the less urgent matters the surrounding crowd has thrust on me. Some of the supplicants I just ignore, but I advise the social worker and the patient impatiently awaiting discharge to wait for the PAs (physician assistants) to arrive. I appeal to the worried mother's understanding and explain that an acutely ill patient needs my help right away, and that I will see her and Hector later on my daily rounds. I ask the nurse to tell Joey S.'s family that I know about the increase in his rectal bleeding and will see them later on my rounds. I coolly advise the obnoxious patient demanding an increase in her methadone dose that I do not discuss such matters at the nursing station and that she must wait until I see her later on rounds. The don't-mess-with-me-this-early-in-the-morning firmness in my voice gives her no chance to reply, and cursing at me in Spanish under her breath, she retreats to her room.

I suspect that Harold M., the kidney-failure patient refusing surgery, is probably playing mind games with me again. He likes the attention his periodic refusals of treatment generate, and I decide to call his bluff. Acting as if I could not care less, I matter-of-factly acknowledge his refusal of surgery and aloofly tell him I will see him later on rounds. Harold is taken aback at my not dropping everything and trying to persuade him to agree to lifesaving surgery, and he warily backs away from the nursing station, probably wondering if maybe he was a bit too hasty in declining surgery. Nothing is more worrisome than having a doctor who seems not to care, but he does not know what I know—that his surgery is scheduled for later in the day, giving him enough time to reconsider his gamesmanship with me.

Finally, for Thomas A., the TB patient who is dressed to go AWOL, I ask the social worker to see him, ostensibly to help him with his welfare check, but really just to buy time until I can see him myself on rounds. And Ciano L., the meningitis patient shooting up drugs, the nurse assures me, is quietly "nodding off" his heroin fix in his room, with stable vital signs and apparently no worse for wear. He can definitely wait for rounds later on.

No sooner are these matters dealt with—or postponed—than a new wave of problems arises, each one thrown at me as an urgent matter that requires my immediate and undivided attention. An X-ray form needs to be completed right away before a patient waiting downstairs in X ray can get his CAT scan done—never mind that I know I filled out the form last Friday. The ward secretary says the hospital pharmacy is on the telephone, undoubtedly to second-guess me, as it routinely does all Spellman physicians, about one of my antibiotic orders. It is not unusual to get a half dozen or so such calls from the cost-conscious pharmacy daily, the majority of them more harassment than helpful advice. Meanwhile, a nurse pushes under my nose a Patient Incident Report Form to be completed about a patient's falling out of bed last night. No harm was done, but the army of hospital "quality assurance" administrators must have documentation that the patient's bump in the middle of the night was really benign. And once again Roberto Q., the patient unhappy about the screwup of his breakfast order, is at the nursing station, loudly carping about what a "dump" St. Clare's is—another satisfied customer, I facetiously say to myself.

Even though I have only been on the unit for ten minutes, I already feel like a little toy monkey strung up on a stick—the kind my parents would buy me at the Ohio State Fair when I was a little boy—bobbing and turning at every tug and jerk of the patients and staff. Resolving that it is too early to feel so overwhelmed, I quickly scrawl the minimum information needed on the X-ray form to get the poor patient through his CAT scan; I ignore the pharmacy's phone call (as I must often do); I nonchalantly throw the Patient Incident Report Form in the wastebasket (my custom under these circumstances); and I look right through Roberto Q., who is now waving his menu request in my face to "prove" he indeed ordered eggs and not oatmeal. Telling him I am a doctor, not a cafeteria employee, would not faze him in the least.

Having beaten back the latest distractions, I retreat into Jose T.'s room, as if it were a refuge from the never-ending bedlam on the unit.

Although a sanctuary from the outside din at the nursing station, Jose T.'s hospital room encloses the more terrifying, more desperate havoc of a forgotten person fighting the awful complications of AIDS. Although I am familiar with the desolation that accompanies this battle, I still marvel at how terrible it is.

Jose's small hospital room is dimly lit, the shadows mercifully obscuring the room's oppressive dreariness—the dirty gray window curtains, the cracked wooden window frames coated with dust and peeling paint, the mud-caked glass windowpanes blocking out the summer sunlight, the nauseating Pepto-Bismol–pink color of the cracking walls, and the detritus of food, human excrement, and blood splattered on the filthy privacy curtain partially drawn around the bed. The sparsely furnished room's only decoration is an anomalous framed and faded theatrical poster announcing *The Heidi Chronicles,* playing at Broadway's Majestic Theater in 1986.

Despite the few furnishings, the air is close, almost claustrophobic. Although the window air conditioner is running, the temperature feels in the eighties. A large fly buzzes around, lands briefly on my pate, then disappears into the closet-sized bathroom. Permeating the room is the familiar smell of stale cigarette smoke, feces, and hospital disinfectant, and as I walk in, my shoes stick to a semicongealed liquid on the floor next to the bed—urine, stool, sputum, spilled juice, or some combination thereof. Strewn on the floor around the bed are clumps of crumpled toilet paper—facial tissues are almost always in short supply at St. Clare's—as well as dried scraps of yesterday's late dinner, a crushed foam cup, some dirty latex exam gloves, and a couple of capsules of unknown medication, which obviously did not make it to their intended destination, Jose's stomach. Cast off in a corner, bundled up in a transparent plastic garbage bag, are Jose's only possessions—prison-issue civilian clothes, a pair of black shoes, and a Yankees baseball cap.

The room's disarray and filth, while typical, nevertheless still impress me: an abandoned ex-prisoner with advanced AIDS has ended up in a hospital room that seems to epitomize both his life and his disease.

Off in the dingiest corner of the room is Jose's hospital bed, hemmed in by a slightly leaning IV pole, a portable tray table (for eating in bed), a wastebasket, an immense oxygen tank on an old metal dolly, a small TV suspended from the adjacent wall by a movable metal arm, and a small bedside nightstand. The chipped Formica top of the bedside stand is stained with dried fruit juice and strewn with scraps of food and empty cracker wrappers. At its edge are perched two white plastic cups. In one are congealed globs of blood-streaked sputum, in the other is cold black coffee in which several cigarette butts float. The open metal drawer of the bedside nightstand reveals a pack of Camel cigarettes, a blue Bic lighter, a couple of folded dollar bills, a partially eaten bag of tortilla chips, and an apparently untouched Bible put there ages ago by the hospital's well-intentioned Pastoral Care staff. The portion of wall closely abutting the bed, like SoHo "splatter art," is decorated with familiar smears—food and juice, human excrement, snot, and a squished cockroach—all familiar because I have seen this montage grow, as if by mitosis, over the span of the last half dozen or so patients who have occupied this bed, my several complaints to housekeeping personnel notwithstanding.

Trying to get closer to the bed, I push aside the paraphernalia around it. The huge, rusted oxygen tank and the dolly cradling it seem to weigh a ton. Prior to coming to St. Clare's, which has no permanently installed oxygen system, I had never seen oxygen dispensed as it was before World War II, but such monstrously sized tanks seem to fit in with the overall antique atmosphere of the place. Moving aside the smeared and stained bedside tray table and its untouched breakfast tray, I am disgusted to feel yet another unknown sticky substance on the edge of the table, and I involuntarily wipe my sticky fingers on the sides of my white coat, which could itself use a good laundering. Cleanliness is always a relative thing on a Spellman AIDS ward, both for patients and for staff.

The only functioning light in the room is a short fluorescent tube on the wall over the head of the bed—the ceiling lights have been out for weeks. I pull the cord to turn on the light and notice that just below it, taped onto the wall over the head of the bed, is the tattered postcard-sized picture of the Blessed Virgin that the mother of the room's previous occupant, Juan H., had put up a few weeks before, in the hope that the Virgin's image would save her son. Juan, however, died last week from pneumonia, after a several-weeks hospitalization. Juan's mother not only bequeathed this portrait of Mary but also left behind the purple rosary beads now hanging from the volume knob of the TV, which is blaring away a *Today* show segment about Mickey Mantle's liver transplant. I turn down the TV volume.

Jose's hospital bed is antiquated, its side rails and rusted metal side frame caked with dried vomit, blood, and feces from several prior patients who had lived and died in it. Although electric, it has not worked for ages, despite the nurses' repeated requests for the maintenance department to fix it. As I get closer to the bed, I am careful not to brush too closely against the side of it, for fear of soiling my pants in whatever fresh grime and gook might be there. A plastic urinal precariously hangs from the bed rail—in it is a slimy mixture of urine, blood-tinged sputum, and more cigarette butts. The wastebasket abuts the side of the bed, and its rim is splattered with brownish sputum, as well as vomit of partially digested food and pills. A small roach scurries over the rim of the basket and disappears under the mattress, into the side frame of the bed.

Momentarily distracted by this stark but splendid tableau of squalor, I have almost overlooked Jose himself, who is quietly lying in bed, amid all of this filth and disarray.

Jose is asleep. A small man—perhaps five and a half feet tall and 130 pounds—he appears only minimally malnourished but is unshaven, disheveled, and unwashed. Dandruff and grease cake his hair, his hands are lined with dirt, and saddest of all, his body odor is the familiar musky, earthy smell of someone who has not bathed in weeks. The chaos that manifests itself in Jose's room extends invariably to the condition within his bed, as if the man and his environment were one. His IV tubing has become disconnected and is silently dripping much-needed IV fluids and antibiotics onto the sheets, which in several areas are already spotted with feces, urine, blood, and juice. Folded into the sheets are cigarette ashes, a few more uningested pills, bits of food (some peas, a grape or two, some crackers), more snot, a few scabs, and many flakes of dried and peeling skin. The IV needle in his wrist is about ready to fall out, since the protective tape and dressing over it have somehow come off and are lying on the floor next to the bed. The plastic oxygen tubing, which should supply oxygen through small nasal prongs, has fallen onto his chin, uselessly oxygenating his lower lip.

Although asleep, Jose appears uncomfortable. His breathing is labored and rapid, and the skin on his face and neck is slightly moist with sweat. He is wearing a badly soiled Mickey Mouse T-shirt from Disney World, and like his bedding, both his T-shirt and bikini-brief shorts are stained with food, fruit juice, IV fluids, urine, blood, and stool. His tanned arms bear silent testimony to a past of danger and violence: the right forearm is decorated with a crude tattoo of a dagger, most likely a symbol of his street gang, and his left arm is covered with old needle tracks from prior IV drug use. Jose has already had a life that I cannot begin to understand.

Jose's condition and his surroundings are as pathetic as they are familiar to me. Words seem inadequate to describe the awful scene before me—the room, the patient, and the disease convey a sense of utter loss. I immediately set about awakening Jose, so I can interview and examine him, to put at least a little order to the chaos I now see. As is often the case with patients on Unit 3A, awakening him is neither easy nor without frustration. My gentle shaking of his shoulder, my quiet calling of his name, indeed awaken him, but severely annoyed at being disturbed, Jose peremptorily tells me, "Hey, man, fuck off . . . leave me alone."

He rolls over in bed, onto his side, and turns his back to me. In so doing, he dislodges his already tenuous IV from his wrist and pushes the urinal off the bed rail and onto the floor below, the slimy contents partially splashing onto my only pair of "office" shoes and missing my pant legs by millimeters. A highly inauspicious beginning to the doctor-patient relationship, I facetiously—and wearily—remark to myself as I step aside from the muck just spilled from the urinal. Now *I* am annoyed with Jose: my first impulse is to tell *him* to fuck off, but I have learned over the years that reacting with anger rarely works with patients like Jose. Such retorts ultimately make my job even more difficult. Indeed, Jose's life, which was never filled with much solace, has up to now been a sadly recurrent refrain of his telling other people to "fuck off," followed by their telling him to do the same. Too many people, among them too many health care workers, do not understand one of the most basic tenets of caring for sick people: namely, sick patients are often not cheery and pleasant. Nonetheless, it is difficult for me always to have to beg, plead, and cajole many of the patients on 3A to allow me to give them medical care.

Apologetically, sweetly, gently, I persist and nudge Jose back awake, telling him how I empathize with his shabby treatment by the prison system and how I must nonetheless take a medical history and do a physical examination if he is to get better. This time his anger abates just a bit, changing to a more complaining attitude, a hopeful sign that he will eventually come around to talking with me.

"Look, man, I'm tired . . . come back later."

"Later is not an option, Mr. T."

Jose then tries the usual protests of a patient who does not want to be bothered—that he is "too sick" to talk to me ("All the more reason to talk to me now rather than later," I reply); that I should instead check the prison medical records, which he brought with him yesterday ("They're practically useless"); and that he had already talked to a doctor in the emergency room ("Yes, but only briefly," I point out). Now, to my relief, Jose and I are already in a dialogue of sorts. Moreover, I have established that he does understand English, an important first step. (Most Hispanic ex-inmates do learn English in prison, if they did not know it before.) As Jose is quickly running out of excuses not to talk with me, I push on, cajole, and negotiate until he reluctantly agrees to talk with me, probably because he is tired of my persistent nagging and perhaps, I hope, because he begins to sense in my approach to him an underlying respect and concern. And although he may not yet realize it, he will eventually experience this same approach from other Spellman staff on Unit 3A.

Having concluded the preliminary negotiations with Jose, I take his medical history and do the physical exam, which are relatively painless for both of us. My persistent prodding yields results: I learn from him that he first tested HIV-positive two years ago in prison, that IV drug use is probably how he contracted HIV, and that his T-cell count was zero earlier in the year. He tells me that he has had no major past complications or hospitalizations as a result of his HIV infection, and that his major symptoms over the past few weeks have been fever, cough, and trouble breathing. Every few minutes or so during the interview, Jose stops to complain about being too tired to continue, but I gently encourage him, putting him back on track with assurances I will hurry along as quickly as possible. Jose is easily distracted and occasionally tries to fiddle with his already cold breakfast tray (I simply push it aside and refocus his attention on my medical questions). At one point, he becomes entranced by the TV program silently playing (I regain his attention simply by turning it off). Knowing that I can press him only so long—his patience is decidedly limited—I zero in on only the most crucial questions about his medical history before going on to the physical exam, which requires only passive cooperation from him. Indeed, during my brief but important interview with Jose, I was already doing part of my physical exam, noting his slow and dull mental status, his mild difficulty breathing (minimal conversation tires him out), and his overall appearance.

Because Jose is too weak to sit up, I have to examine him as he lies in bed. He is barely cooperative and acts put out and as if I am wasting his time. My physical exam turns up nothing too surprising for an ex-inmate with AIDS. I find mild generalized muscle wasting, an old knife wound on the right side of his face, terrible oral hygiene (rotting teeth caked with dried plaque and old bits of food), thrush (the white fungal coating on the tongue and back of the throat), stertorous lung sounds, a healed midabdominal surgical scar from an old gunshot wound, more old needle tracks on his legs (which the bedsheets previously covered up), and another fading tattoo—the word *Madre* ornately inscribed on the upper part of his back. This tattoo, apparently a remote tribute to his mother, reminds me to ask Jose about any close family members in the New York area, primarily to see if he wants any of them contacted about his admission to St. Clare's. As is often the case with 3A patients, there are none, and he says he has not spoken with his mother and sisters in Puerto Rico for twelve years. I have heard even more desolate family stories on Unit 3A, but I quietly wonder and marvel at how someone could at one time in the past have felt so deeply about his mother that he would tattoo her memory on his back and yet later not communicate with her for twelve years.

My physical exam completed, I am heartened that my visit with Jose has gone so well. Not only did he acquiesce with minimal protest to the interview and exam, but my visit with him has been uninterrupted. Usually during my rounds, a distracting procession of other hospital staff parades in and out of the patient's room—nurses with medications, aides with food trays or clean bedding, respiratory therapists changing the behemoth oxygen tanks, or housekeepers to empty the trash baskets. Although I always ignore the intrusions and go on with my business with the patient, the interruptions can break my train of thought, sometimes causing me to overlook important matters. My patient visits always seem to be cut short by my beeper announcing a phone call from outside the hospital—and I never can predict whether such an outside call is important enough to interrupt my visit for the call. But so far today Jose and I have been lucky.

Now that Jose's interview and exam are finished with a minimum of fuss, I set out to explain to him my diagnostic impression and, most importantly, to enlist his cooperation in his treatment. Although most of my patients on Unit 3A are largely uneducated, they still have a rough idea of what AIDS is, and although largely passive and uninterested in my interview and exam, Jose indicates that he, too, does not really need to hear my one-minute "HIV Disease Made Simple" lecture that I give to those few new patients who do not have a clue about AIDS. With Jose, I dispense with the pathophysiology speech and move on to tell him that his AIDS has caused him to get a pneumonia. This pneumonia could be the garden-variety bacterial pneumonia or TB or—more seriously—it could be PCP, *Pneumocystis carinii* pneumonia, "the AIDS pneumonia." Jose seems to understand so far what I am telling him. Many inmates and ex-inmates with AIDS have seen their fellow prisoners fall ill with AIDS-related problems, and PCP—as well as TB—are not unfamiliar to such patients, who otherwise might not even know, or care about, other current events in the world.

Jose listens with a passive stare on his face, and I continue on and explain that a battery of tests and procedures are needed to help determine exactly which type of pneumonia he has. I briefly review these tests for him: the two most important tests are a bronchoscopy ("We give you a tranquilizer and put a small telescope into your windpipe to look at your lungs and get a biopsy to see if you have PCP") and collecting three sputum specimens for TB ("Cough into the cups the nurse gives you"). I warn Jose that until these three sputum samples return negative for TB, he must stay in his room and not go out in the hallway or the patient lounge—that is, he must remain in "respiratory isolation." To help him deal with such isolation, the Spellman Center will provide a free telephone (incoming calls only) and a free TV. Finally, I inform him that, until the tests diagnose his type of pneumonia, he will be treated with intravenous Bactrim, which will cover both PCP and bacterial pneumonia.

I deliver my speech fairly quickly and ask Jose if he has any questions. He nods in the negative. I still think he understood most of what I hurriedly threw at him just now, but I am not sure. It would be nice, I think to myself for the thousandth time at St. Clare's, if I could really spend more time explaining things to patients such as Jose, but there is never enough time for even the bare essentials on Unit 3A. So my brief discussion will have to suffice for now. Besides, Jose is probably not in the mood for an extended and detailed analysis of his condition.

Finally, because Jose is so sick from pneumonia—there is no guarantee he will improve with treatment—I feel obliged, before I leave, to ask if he has given any thought to the important issue of cardiopulmonary resuscitation (CPR). That is, I ask, what—if anything—does he want to be done if and when he is about to stop breathing or his heart stops beating? Does he want us to pound and pump on his chest to try to get his heart working again? Does he want us to put a tube down his throat, into his windpipe, to hook him onto a breathing machine in the intensive care unit? Or, if and when he reaches the threshold of death, would he prefer just to be kept comfortable and "let nature take its course?" To conclude this part of my CPR speech, which is delivered with alacrity, I quickly tell Jose that New York State law stipulates that, unless a patient signs a Do Not Resuscitate (DNR) form, every in-hospital cardiopulmonary arrest *must* immediately be treated with a full-scale resuscitation attempt, regardless of the chances of success.

As I have so often with other patients, I am wishing I had more time to review such important life-and-death issues with Jose. My own mind, however, is now wandering off to my full day of chores—my other sixteen patients and their special problems I have yet to confront today, as well as the family members waiting to talk to me about their loved ones; the hours of paperwork and note-writing in patients' charts; and the plethora of disputes, problems, and decisions that will need my undivided attention. Glancing quickly at my wristwatch, I repeat my question to Jose a bit differently, but still very matter-of-factly: Does he want to sign a DNR form, does he want "everything possible" done to keep him alive, or does he want to think about it and let me know later?

Jose does not answer my question directly, but as sometimes happens at moments like this, his reaction to this delicate subject is revealing. Although uneducated, Jose has seen enough movies and TV to know what life-support systems, breathing machines, and intensive care units mean. His indifference suddenly disappears, replaced by an acute, genuine concern. This pain-in-the-ass doctor, whose pokings and proddings disturbed his first good sleep in many days, is now talking about death—not death as an abstraction, but the death of Jose T. At last, I have gotten Jose's attention.

"I'm really *that* sick?" he fixedly asks me. "It's not *that* bad, is it?" And, of course, the ultimate question that is both simple yet profound in its implications: "I'm not going to die, am I?"

Suddenly Jose is facing the fact that AIDS kills, and he is palpably terrified, his face and voice reflecting a fear I have seen many times in patients on 3A. I know that I am facing a doctor-patient moment that is as needful of my precious time as anything else awaiting me on 3A that day.

Jose's urgent questions—his sudden animation and fear—give me the first chance to make an emotional connection with him. And I must make this crucial connection and build upon it if I am ever to have Jose's cooperation in his medical care and—more importantly—if I am ever to help him face his disease. These important opportunities to reach a patient's soul come at different and unexpected times—sometimes, as today, when the specter of CPR is brought up for discussion, sometimes after weeks or even months of power struggles with a patient, other times right before death, and sadly too often, never at all.

Consequently, Jose's sudden realization of his mortality and his urgent need for reassurance crystallize an otherwise routine first visit into an opportunity for both of us to go beyond the mechanics of AIDS medicine—to establish a humane dimension to our previously sterile interaction. I reach over and gently touch his hand. Although I have physically touched Jose many times during the prior medical examination, this gesture somehow conveys a concern he has probably not felt in a long time.

Looking directly into his eyes, I try to give him both the truth and reassurance. Yes, I tell him, he is very sick and could possibly die from the pneumonia . . . *but,* I quickly add, it is also very possible he will respond to treatment, especially if the two of us work together and not against one another. I promise always to be honest with him and to give him my best care and attention—promises most middle-class patients would take for granted but which might not seem so evident to Jose, especially in light of his recent medical care in prison. As for the CPR/DNR matter, I assure him that he does not need to give me an answer now, and that if he eventually opts to sign a DNR form, he is neither "throwing in the towel" nor signing away any medical care other than resuscitative measures right at the time of death. I further try to lessen the enormity of the CPR issue by observing that such life-and-death issues should be addressed by everyone, regardless of HIV status. "I could get hit by a bus later today and face death before you do," I point out to him. And finally with regards to death—and here I squeeze his hand a little tighter, to keep his attention and to reassure him—I promise to do everything possible to ease his suffering, when it comes.

Jose has listened closely to my reassurances and appears less anxious. After a brief silence, he says he wants to be kept alive, "no matter what," at least until he has had a chance to see his mother, the mother to whom he has not spoken for twelve years. As he mentions his mother again, his eyes become slightly teary and his voice cracks with sadness. "She might be dead by now, or maybe she thinks *I'm* dead. . . . I don't even know how to reach her. . . . I don't know where to start looking. . . . I should have never kicked her out of my life, but the gangs made me do it." I touch him on the shoulder and explain that locating his mother is exactly the kind of challenge 3A's social worker relishes. "If anyone can find your mother, our social worker, Abby A., can do it—she's a magician."

For several more seconds, my thoughts remain on Jose's long-lost mother and whether she would even want to know about her son, who had apparently cut her out of his life twelve years ago. But knowing most mothers—and recalling many similar instances of prodigal sons reaching out to their once-spurned mothers for solace from AIDS—I reckon that Jose's mother would be transfixed with joy to hear from him. As I think about Jose's mother and the connection I had just made with him, an indescribable sensation of awe and gratitude overcomes me. Jose T.'s life of drugs and gang violence has progressed to his present lonely struggle against a disease that few knew about twelve years ago, when he last spoke with his mother, the only woman who has remained a constant in his life. And amid the unspeakable squalor of his hospital room, and in the setting of a horrible disease that could, I suspect, well be a metaphor for his life itself, I am able to feel a part of his anguish, thereby narrowing the seemingly insurmountable gulf between the two of us.

My musings are ended abruptly by my beeper, which announces an out-of-hospital telephone call. Moreover, the tumult of Unit 3A awaits.

Completely exhausted, Jose has no further questions, and excusing myself, I step into his bathroom to wash my hands, pleased with how well our first visit went. My satisfaction is quickly deflated—replaced with more typical exasperation—when I find that his bathroom, as is often the case on 3A, has no soap or paper towels. But Jose's frail but sincere "Thanks, Doc," as I leave his room, rings beautifully in my ears, reminding me of the light-years of progress made since his first volley of "fuck off" greeted me thirty minutes earlier.

Emerging from the tranquillity of Jose's room, I am once again part of the maelstrom of the Spellman menagerie. Dozens of different and competing patient needs and demands occur simultaneously, without regard for civility or rationality.

One of my patients, a homeless IV drug user admitted for a large arm abscess acquired by shooting up (and missing the vein), is camped out at the nursing station, with IV pole at his side. He is barefoot, and his scanty, untied hospital gown is open in the back, exposing his bare back and woefully unwashed underwear. He is rudely badgering a nurse, who is on the phone, about getting his pain medicine *now,* despite the fact his next dose is not due for another two hours.

"I'm in a lotta pain . . . ya gotta give me my Percocet," he whines, almost tilting over the wobbly IV pole as he leans on it for support.

Trying to ignore the patient's entreaties, the harried nurse is arguing with the dietary department about Orlando V.'s breakfast tray, which was not sent up to the unit. Dietary says it has been, but the nurse says it *most definitely* was not. Just now Orlando, who has TB and is not renowned for patience—he has been waiting in respiratory isolation for almost two weeks for his sputum smears to return from the lab—swings open the door of his respiratory-isolation room and, spewing his germs into the busy hallway, yells to his nurse that he is walking out to a corner deli for food if his breakfast does not arrive in five minutes.

"You're all trying to fucking starve me!" Orlando bellows.

Another nurse farther down the hallway sees me and calls out that Narvor H., an extremely sick patient with both meningitis and a brain tumor, is refusing to go down to X ray for a repeat head CAT scan. Moreover, throughout the weekend he has been disconnecting his IV antibiotics whenever the nursing staff would hang them. I suspect it is Narvor's dementia and confusion, not his chronic antisocial personality disorder, that is to blame.

The unit's respiratory therapist is also at the nursing station and, seeing me emerge from Jose T.'s room, barely waits for the first pause in the above salvos and commences to gripe to me how Mrs. Joy D., who is on oxygen for pneumonia, is still smoking cigarettes in bed, despite multiple confiscations of her cigarettes and matches by the nurses over the past week. And adding to the overall confusion, a housekeeping employee is waxing the hallway floor with an immense, noisy buffing machine, which is causing major gridlock for the several laundry hampers and medication carts the other nurses are trying to maneuver through the narrow hallway. The floor wax has an overpowering, nauseating odor that wafts through the hot unit and makes the noise and multiple patient problems even more annoying. The brief euphoria of my visit with Jose T. has dissipated.

My beeper goes off again—the outside phone call is impatiently waiting. Happy to duck the insoluble problems confronting me at the nursing station, I run for a phone to pick up the call, hoping it is not yet another crisis. But, in fact, it is. Mrs. C. is trying unsuccessfully to care for her thirty-five-year-old son, Ricardo, at her home. Ricardo was just discharged from my floor last week, after a protracted four-month stay for multiple, intractable complications of his HIV disease. At his discharge, Ricardo was so weak and emaciated that he could not even feed himself, and over the prior weeks, I had repeatedly advised both him and his mother to consider hospice nursing-home care. But, as is often the case, they would hear nothing of going to a nursing home and insisted on a try at home care, which was arranged with around-the-clock nurses and home health-care aides. Yet Mrs. C. reports over the phone that Ricardo is now even weaker, in intense pain from chronic HIV nerve damage in his feet, and unable to eat or take his medicines.

Despite the craziness of 3A still swirling around me—a nurse is patiently waiting beside me to report yet another problem as soon as I hang up, and Orlando V., the starving TB patient, is also out in the hallway and literally in my face, yelling about his undelivered breakfast tray—I can hear the desperation in Mrs. C.'s voice as she tearfully asks me what I can do to help her sick son. Sometimes a doctor's job is to tell people the obvious, even if they do not want to hear it. I patiently tell Mrs. C. that she must bring Ricardo back into the hospital at once. What I do not tell her is that this time he will almost inevitably have to be discharged to a nursing home. That is an issue for much later, *if* he makes it out of the hospital alive this time.

As was often the case on Ricardo's prior admission, my advice—that he be readmitted immediately—is not what Mrs. C. wants to hear. She ignores my advice and starts grasping for straws of hope: maybe Ricardo needs a "vitamin B shot," she proffers, or maybe a softer mattress will ease his pain. I refuse to offer her false hope, especially this late in Ricardo's disease: he is very, very sick, I firmly tell her, and must definitely be in the hospital—*now.*

Yet Mrs. C. stubbornly perseveres with her aimless attempts to dodge the real issues: she complains about the home health nursing care, wonders out loud whether a softer air-mattress might help Ricardo's bedsores, and asks if there are any stronger pills to stimulate his appetite. I likewise ignore her questions and comments and simply reiterate my only advice: bring Ricardo to St. Clare's *now.*

After several more minutes of Mrs. C.'s rambling commentary about the most minute details of Ricardo's incapacitation—she has always been persistent in her denial of how sick Ricardo has been—she runs out of excuses and reluctantly acquiesces to calling an ambulance to bring Ricardo back to St. Clare's. Our conversation ended, I immediately call the admitting office to reserve for Ricardo the 3A bed just emptied by Mr. Y.'s discharge.

It is only nine—I have been on 3A for only an hour—but the enormity, the *impossibility,* of the work yet to be done sinks in. The nurse waiting for me to finish my phone conversation with Mrs. C. now reports that she suspects Mr. Roberto Q.—the patient upset earlier about getting oatmeal instead of eggs—is not taking his TB medications, because she found several days' supply of pills hidden under his mattress when she changed his sheets this morning. Astonishingly, despite the nurses' carefully observing TB patients taking their medications, many patients are wizards at deceiving even the best nurses by secreting unwanted pills under the tongue or inside a cheek, disgorging them later. Roberto is one of the few patients I thought would not pull such a trick—so much for my judgment of character, I think to myself—and I make a mental note to confront him about this matter later on my rounds. Both the nurses and I know that all of us would have gone crazy a long time ago if we reacted too indignantly every time a 3A patient refused medication or X rays or other lab tests.

Seeking out a quiet space on the unit, I settle for the corner next to the laundry chute. The relentless buffing machine has already waxed the floor there, so I figure that only the aides periodically emptying laundry down the chute will disturb me. I grab a chair from the nursing station, position myself in the corner as inconspicuously as possible—back turned to the rest of the unit to disinvite any interruptions—and begin to write out, in longhand, Jose T.'s medical history and exam, revising his medication orders and ordering some blood and X-ray tests. The miracles of "modern medicine" notwithstanding, the medical chart paperwork takes much longer than the time I actually spend with the patient.

As almost always happens, my frantic scribblings are interrupted. First another out-of-hospital telephone call from a visiting nurse asks for clarification of a medication order for a patient discharged from 3A a few weeks ago. Then one of the hospital pharmacists calls to ask if I really think it necessary to treat one of my patients with antibiotics for two weeks instead of ten days (yes, I really do!). Spying me off in the corner, the Spellman discharge planner requests my signature on some nursing home applications that need to be submitted right away. I take several more outside calls from patients' family members wanting to know how their loved ones are doing (as far as I know, they're okay—on 3A, no news from the nurses about a patient's condition is usually good news—but call back later in the day, please, after I have had a chance to complete my rounds). And finally, I receive an outside call from a physician recruiter wanting to know if I would be interested in a "lucrative medical practice opportunity" in beautiful rural Missouri, "rich in outdoor recreational activities and only an hour's drive from St. Louis, with all its cultural attractions" (no thanks—Central Park is outdoors enough for me, and I live a ten-minute walk from Lincoln Center). After Jose T.'s paperwork is finally finished, I briefly go over his case with the 3A physician assistant who will be following him along with me. To finish his care for today, I call the unit's social worker, Abby A.—a "plainclothes" nun and a supremely dedicated social worker and good friend—and ask her help in locating Jose's long-lost mother in Puerto Rico.

By now things seem to have quieted down on 3A. Orlando V. has finally gotten his breakfast tray from the hospital kitchen, but because the food was cold, he threw the tray into the hallway and indignantly stormed out to a nearby deli to fetch an egg sandwich and coffee, returning about twenty minutes later. The meningitis patient caught earlier shooting heroin in another ward is still nodding off his fix, quite stable according to my PA. The other PA took care of the discharge paperwork for Mr. Y., who, so impatient earlier about being released first thing, is now all smiles as he thanks me on the way out the door. The other patients have gotten their methadone and pain medicines and are likewise nodding off in their rooms, mercifully out of everyone's way. And the relentless waxing and buffing of the hallway floor has been completed, just in time for one of the nurse's aides—a short and inescapably rotund lady—to slip on the slick floor and drop a full bedpan all over the area in front of the elevator, thankfully without hurting herself seriously.

Feeling a little less hassled, I set out with my two PAs to start daily rounds on the other patients on the unit, but I have the uneasy feeling that I have forgotten something—or someone—I had earlier promised to take care of. I cannot quite remember, however, what this important something or someone is. But my memory is immediately jolted when, just as my PAs and I are about to begin rounds, a nurse frantically approaches us to report that Hector R.—the patient whose worried mother I had earlier in the morning promised to talk with right after I took care of Jose T.—has just been found "unresponsive," without pulse or respirations.

This is a moment I have been dreading for weeks.

Hector R. is a twenty-eight-year-old man, a former state prisoner, with AIDS-related lymphoma, a rapidly progressive cancer of the lymph glands that in AIDS cases almost always means death within four to six months after the cancer is diagnosed. Although some patients might respond to chemotherapy for a few extra months, the tumor quickly relapses, with subsequent deterioration and death. Admitted to St. Clare's almost three months ago, Hector was at that time judged too weak and emaciated to tolerate even reduced-dosage chemotherapy for his lymphoma. Although he was a state prisoner when he was admitted, Hector's AIDS was so advanced that he was quickly given a "medical parole," a parole that meant little, since he was then simply transferred from the much cleaner and quieter Spellman prison ward to the more noisome ward on 3A, where I took over his medical care.

By the time Hector reached my unit a month ago, his debility and AIDS-related dementia were so severe that he was unable to cooperate with his care. He was constantly pulling out his IVs, would not eat or take most of his medications, and repeatedly pulled out the nasogastric feeding tube, which had been inserted at his mother's insistence because of his inability to eat. Hector's dementia, caused by both HIV and the lymphoma's spread to his brain, gradually worsened to the point that he had no idea what was going on around him. Staring blankly into the air, he would appear awake but could only tell me his name, and over the past week or so, his deterioration progressed to the point that he sometimes could not even remember that. Accordingly, Hector's mother, his only close family member, was responsible for making medical decisions for him.

Mrs. R. was a gentle, self-effacing lady who seemed totally overwhelmed by the enormity of what was happening both to her son *and* to herself. She would faithfully visit Hector every day, always bringing in food he would never eat and gifts he could never appreciate. As Hector's condition worsened, she was allowed to stay beyond regular visiting hours, and recently she would remain overnight in his room, sleeping fully clothed—a Bible resting open on her lap—on an uncomfortable lounge chair. Mrs. R. had already lost two older sons to AIDS—both had died in prison and, like Hector, had contracted HIV via IV drug use. Estranged for decades from her husband and living alone, she often appeared bewildered and forlorn, and she devoted her entire energies to her last son, helping the nurses with his care, cleaning his room by herself, and always talking to him, even though it was obvious his profound dementia blocked out most of what she would say to him.

For the past month I would speak with Mrs. R. almost daily about Hector's condition and would repeatedly raise with her the issue of CPR, the same issue I had raised earlier this morning with Jose T. Mrs. R. would always demur—she decided not to decide, leaving the matter to "God's will." As I often did, I enlisted the help of Sister Pascal in trying to convince Mrs. R. to agree to a DNR—Do Not Resuscitate—order. But even Pascal's supportive words and advice had no effect. Mrs. R. felt that after seven years in prison, Hector deserved to be at home with her, and she did not want to hear anything to suggest that this overriding goal was probably not attainable. As Hector's medical complications multiplied—recurrent pneumonias, worsening and enlarging bedsores, and serious blood infections—plans to discharge him to his mother's for twenty-four-hour nursing care had to be repeatedly postponed.

Yet with each setback in Hector's condition, with each postponement of Hector's long-hoped-for discharge to home, Mrs. R. remained undeterred, and even encouraged, in her optimism that "God's will" would soon become manifest. As Hector gradually wasted away to a virtual skeleton and developed his third pneumonia in as many months, she tenaciously clung to hope for "a miracle," refusing to listen to my gloomy pronouncements about the utter hopelessness of her son's condition. Pastoral care staff, Abby A., and the nurses also tried to help Mrs. R. accept Hector's impending death, without success. She wanted "everything possible" done to keep him alive, so he could eventually come home with her.

"He's all I have," she would explain, "and I want to be there at his side when he finally goes home to the Lord. . . . It almost killed me when I couldn't be with my other boys when they passed away in prison, and I want to be there with Hector."

My frustration with Mrs. R.'s denial about Hector's terminal prognosis was short-lived. As an AIDS doctor, I had dealt with many similar cases of denial by both patient and family and had eventually come to respect it as a sometimes necessary coping mechanism, an emotional cushion against realities too painful to bear. Years earlier in my career, I had been less charitable in my opinion of patients and family who refused to face death head-on, but my experience at St. Clare's convinced me that death and suffering are probably no different when dealt with realistically as opposed to in denial. As Hector's condition spiraled downward, I eventually reached the point where, in a medical sense, I really did not worry about his case at all—I was only concerned that he not have any pain or discomfort. My focus on Hector shifted almost exclusively to helping prepare his mother for the grievous hurt his death would cause her. As so often happens on my AIDS ward, I began worrying more about the patient's family than the patient himself.

Now, the nurse's alarming report that Hector is unresponsive means that the final moment has come for both Hector and his mother. "God's will" is about to be fulfilled.

As my two PAs and several nurses briskly walk with me to Hector's room, there is no question about whether or not to try to resuscitate him. New York State law is quite explicit on the matter: unless the patient or the patient's health-care proxy has signed a DNR form, cardiopulmonary resuscitation *must* be attempted, regardless of how bleak the chances of survival might be. As with so many other health-care issues, this legal imperative to try to revive every hospitalized patient who is about to die—most patients in New York have no idea what the DNR issue is all about—this legal imperative is an area where the law is an ass, and I am loathe to use a more tasteful word.

DNR is a matter about which a doctor is wise to be a tad paranoid. An underemployed trial lawyer or a publicity-hungry assistant district attorney or a doctor-hating hospital employee could ruin a physician's career by claiming lifesaving CPR was illegally withheld from a patient like Hector R. Sometimes—when a terminal, hopeless patient has no family—I, like all of the other Spellman physicians, would bend the law by only going through the motions of CPR in what would cynically be termed a "slow code," where the chest compressions would be flimsy, the artificial breathing for the patient would be shallow, and the overall pace of the code would be formal and glacially slow. CPR's otherwise frenetic and often violent "dance around the bed," as I cynically call it, would be more like a pavane or saraband. Such a slow code would satisfy the letter of the law but not its spirit, since the ultimate goal would be to let the patient die with a minimum of medically sanctioned violence to the body—and a minimum of inhumane indignity to the spirit. I hate the supreme hypocrisy of a slow code, since its emphasis on form over substance somehow represents all that is wrong with modern American medicine and its subservience to inane legalisms.

Hector's CPR cannot be finessed into such a merciful slow code: Mrs. R.'s insistence on doing "everything possible" for her son compels me to follow the letter of the law rigorously. I am really too tired at this point in the day to be angry about what the staff and I are about to do to Hector—indeed, mightily have I tried to let this cup pass from him. Yet I still worry about how his mother is going to take his impending death. As my team and I approach Hector's room, one of the nurses tells me that Mrs. R. left the floor just twenty minutes ago, most likely to get some breakfast at a nearby coffee shop. It *always* seems to happen this way, I think to myself. As soon as a devoted family member leaves for just a few minutes and breaks the interminable bedside vigil for a terminally ill loved one, the patient dies. It is almost as if the patient, even when completely comatose, wants to end his or her suffering alone and with silent dignity, away from the sad gazes and misguided pity of the living. However, as mandated by state law, we are not about to let Hector leave this world either alone or with silent dignity.

Hector's room and everything in it are clean and spotless—Mrs. R. had seen to that. The small bedside TV is blaring away with an obnoxiously loud, confrontational morning talk show. Mrs. R. always tuned in the TV for Hector, even though he was never watching it. Normally I would promptly turn down the TV volume when visiting a patient, especially for something as noisy as CPR, but this time I leave the TV as is. feeling that the stridency of the tube will be perfect background accompaniment for the violence about to be played out in the room.

As feared, Hector's skeletal body is inert, without pulse or respirations, but is still warm to the touch, indicating his unresponsiveness has been present for a matter of minutes. Hector's malnutrition has been so profound over the past months that he looks like a victim of the Nazi camps, or a faceless casualty of a sub-Saharan famine. Nonetheless, the code cart—replete with its high-tech paraphernalia for resurrecting the dead—is dutifully wheeled into the room, and everyone sets about this gruesome dance macabre around Hector's deathbed. One PA puts a bag mask over Hector's sore-encrusted nose and mouth and attempts to force oxygen-enriched air into his now airless lungs. The other PA, a muscular rock climber, vigorously pumps up and down on Hector's frail chest, trying to pump blood through what little remains of his body. I myself am more detached from the action and watch the heart monitor that the nurses have just applied to Hector's chest, periodically giving intravenous drugs to try to restore his heartbeat. The nurses methodically draw up the intravenous medications and carefully record on his chart whatever is done or given to him. Always summoned to such codes, Father Jack arrives a few minutes later and, standing at the head of the bed, unobtrusively administers the Last Rites. Sister Pascal stands guard outside the doorway, simultaneously praying and being on the lookout for Mrs. R., to intercept her, lest she return from breakfast too soon and see what is being done to her youngest and only remaining child.

As with most codes, Hector's is not a pretty sight. The necessary forcefulness of the chest compressions breaks both his protuberant breastbone and several of his brittle ribs, resulting in a sickening crunching sound as the PA pumps on his chest. I recall one of my physician mentors during internship reassured me that cracked ribs during CPR were good evidence that the chest compressions were of adequate force to pump the heart. The chest pumping is so vigorous, and Hector is so small, that his body appears to convulsively jerk about in the bed with each downward thrust the PA inflicts. Eventually, every chest compression becomes transmitted to Hector's lower abdomen and forces out more and more liquid stool into an enlarging yellow puddle between his thin, bony legs. The air being forced into his lungs also distends his stomach, causing dark green vomit to gurgle up through his nose and mouth. Soon, despite attempts to suction the vomit out of his mouth, a green pool of puke encircles his head and mats his thin black hair. Hector's eyes remain open but vacant, the chest compressions giving his glassy stare a wide, bug-eyed appearance. The TV screams out a raucous talk show about men who have left their overweight wives for thinner women—enraged fat women in the audience are yelling insults at smug husbands onstage, the bedlam interspersed with ads for furniture wax and ambulance-chasing lawyers.

The crush of people and equipment around Hector's bed creates a gridlock of IV poles, oxygen tank, code cart, EKG machine, suction tray, and bedside tables. Someone accidentally knocks onto the floor the small radio and vase of flowers his mother brought in for his birthday the previous week. Except for necessary communication, everyone in the room is grimly silent, quietly hoping that Hector will stay dead. All of us know that even if his heartbeat could be restarted, he would still have to hooked up to a breathing machine in intensive care, where, given his profound malnutrition, he would linger on for no more than a few days, eventually dying tethered to the inhumane hardware of high-tech ICU medicine.

After about fifteen minutes of CPR, Hector is still dead, and I call an end to this obscene farce, New York State law having been followed to the letter. The PAs, appearing like Charon's attendants, are allowed to stop their poundings and pumpings, the underworld having been reached. The flatline heart monitor is turned off, and the nurses begin methodically cleaning up the mess. The pretty blue handmade quilt and matching down pillow his mother had brought in long ago are badly stained with blood, vomit, and stool. Pushed onto the floor by the chaos of the CPR are the two little stuffed teddy bears that were always in bed with him. Taped onto the wall, over the head of the bed, are a set of rosary beads, a small picture of Jesus, and several old and tattered snapshots of a much younger Hector with his mother and two older brothers in happier times past, before drugs, prison, and AIDS destroyed their family. The TV is still loud. I turn it off.

I look at Hector and know that I must not allow myself to feel either sadness or indifference. Given the sheer magnitude of death and suffering on 3A every day, either emotion every time a patient like Hector R. dies would destroy me. But as I pause again to gaze on the memorabilia in his room, I am overwhelmed with an acknowledgment of life's preciousness.

After thanking the staff for their help, I go into the hallway to see if Hector's mother has returned from breakfast. Sister Pascal tells me that Mrs. R. had come back as the CPR was in progress. Fortunately Pascal kept her from seeing what was being done, and Father Jack has taken her upstairs to the pastoral care office, to wait for word on what was happening to her last son. As both Pascal and I walk toward the pastoral care office, she remarks how wonderful Hector's care has been on 3A and rejoices that Hector "had the good sense to die," despite his mother's misguided demands he be kept alive at all cost.

"Hector has gone on to a far better place," Pascal reassures me with her loving, matter-of-fact aplomb. Always my biggest booster, Pascal knows when a case such as Hector's has bothered me, and her accepting approach to dying carries a joy that manages to lift my sagging spirits.

In the small but cozy pastoral care office, Mrs. R. is sitting next to Father Jack on the sofa, quietly sobbing—a good sign, I think to myself, since it means she has been expecting the worst. She sees Pascal and me enter and reads our faces of condolence—and she instantly knows that her son is dead.

After almost two decades of announcing death to people like Mrs. R., I still feel that the direct approach is best.

I reach down, touch her hand, and with quiet sadness say, "I'm sorry, Mrs. R., but Hector is dead."

She screams in anguish and grief, but fortunately is not hysterical. As expected, Mrs. R. blames herself for leaving Hector to go for breakfast.

"I shouldn't have gone, I should have been there for him. God only knows how much I wanted to be there with him when he passed. If I had been there to help him, this wouldn't have happened!" Her emotional pain seems unbearable.

Mrs. R.'s self-blame mobilizes me into action for my final job as Hector's doctor: to lessen his mother's suffering. Invoking all the moral authority I can as "the doctor," I gently command her to stop crying, to look directly at me, and to listen carefully to what I am about to say: that even though she was not at his bedside at the moment of death, her love always was and always will be there for Hector . . . that her love and presence were the very last things Hector experienced before he died . . . that it was her love, not medical science, that comforted Hector and mattered the most to him . . . that he did not suffer at death—death was without pain . . . and, most important, that the love between her and Hector did not die in his room just a few minutes ago.

Forcefully, but respectfully, I repeat these admonitions several times to her, holding her hands and firmly insisting on eye contact through the tears so she can concentrate on what I am saying and not be so distracted by the overwhelming grief. My words of comfort are intended as a loving rebuke—I admonish her that she should never minimize the profound effect her ever-attentive presence had on her son. After I am fairly certain she understands how her self-blame unfairly diminishes her love for Hector, I close by telling her of the fondness all of the 3A staff had for Hector and of the sadness everyone on the unit feels at his death. Finally, I encourage her to feel free to call me anytime in the future she has any questions or concerns about Hector's care.

Mrs. R. seems relieved by my comments, although by now she appears completely exhausted, emotionally and physically. She thanks me profusely even though I feel such thanks are undeserved, and I leave her to Pascal and Father Jack, who will take her to Hector's room in a few minutes, so she can see him one last time in his hospital bed.

Hector's CPR has set me behind schedule: I still have a mountain of work to finish on the ward. As I hurry back to 3A to resume my rounds, I pass by Hector's room and pause briefly to peek in. The nurses have already picked up the mess from the CPR, and Hector and his personal effects have also quickly been cleaned up—the vomit, blood, and stool have been wiped away, so his mother will not be too shocked when she stops by shortly for a final viewing, before he is zipped up into a body bag and shipped down to the morgue. Gazing from the doorway at Hector for the last time, I reflect on how I wish I had more time to spend with Mrs. R. at her son's bedside in a few minutes, but there is no more time to spare. An ancient rabbi once said of death, "Let the dead bury the dead," but he would undoubtedly have tried to comfort the bereaved living, such as Mrs. R.

Several hours later, after I have finished my rounds, I am again seated in my little private corner near the laundry chute, frantically writing in patients' charts. I look up and spy Mrs. R. waiting at the 3A elevator for the last time, holding a large plastic bag in which she has collected Hector's belongings—the soft pillows, the teddy bears, the pretty quilt blanket, the family pictures, and all the other little things that she had brought in for him over the prior months. Not knowing she is being watched, she appears frightened, very much alone. I despair at my utter helplessness: I know there is absolutely nothing more I can say or do to comfort her.

After Hector's CPR, my PAs and I once again set out on patient rounds, and the rest of the day speeds past.

The first patient on rounds is a problem left over from the first onslaught of the morning. Mr. Thomas A., the TB patient who was threatening to leave to pick up his welfare check, is still in his street clothes and has gathered his things into a plastic bag, fully intending to go AWOL. Abby A. saw Thomas earlier but was unable to help him get his check (by law, only the actual recipient can pick up his or her welfare check). Both Abby and I knew earlier this morning that there was probably no way to help him in this matter, but her visit bought crucial time while I was busy with Jose T. and Hector R. Fully aware of the public health risk Thomas's elopement would create and knowing I cannot physically keep him in his room against his will, I first try on rounds today my usual routine of negotiating, begging, and cajoling to keep him on the ward. I offer him methadone or tranquilizers such as Ativan or Xanax as enticements to stay put, since it is always possible that he intends to use his welfare money to buy drugs, as he has done in the past. But Thomas is undeterred and begins to collect his few belongings on his way out the door. I thus have no choice but to play my trump card. Because Thomas is an ex-prisoner on parole, I coolly remind him that his leaving the hospital while still infectious with TB will force me to notify his parole officer. This will cause his parole to be revoked, sending him back up the Hudson River to Sing Sing. Checkmated, he loudly curses me and throws his bag of possessions back onto the bed. Always one to turn a colorful phrase—he was once a Marine Corps sergeant—he snarls, "You think you're so damn smart, but let me tell you: the caveman forgot more than you'll ever learn!" Although the ACLU might quarrel with the bluntness of my threat to Thomas, I am pleased with myself for keeping him off the streets of New York.

Next on rounds is Ciano L., the meningitis patient who had earlier wandered off 3A to shoot up drugs on another floor of the hospital. One of the city's many homeless drug addicts—he prefers the train tunnels of the Times Square subway stops—Ciano has been a major "problem patient" in the past. On a recent admission for pneumonia, he had hidden in his bedsheets his drug paraphernalia—an old beat-up syringe and needle—and a nurse stuck herself on the needle while changing his bed. For that infraction, he was kicked out of the hospital in what was termed an "administrative discharge"—he had by that time sufficiently recovered from the pneumonia—and burly hospital security officers came to his room, stuffed his meager personal effects into a plastic garbage bag, and escorted him out of the hospital, leaving him and his garbage bag at the corner of Ninth Avenue and Fifty-first Street.

This time, however, Ciano's fungal meningitis makes him far too sick to be kicked out again. He would quickly resurface in the emergency room and have to be readmitted, causing unnecessary duplication of work for everyone. Thus, today I negotiate with him an increase in his methadone dose, in return for his solemn promise to stay in his room. Better that he be snowed under by methadone than have him hosting heroin parties off the unit.

The next patient on rounds, Elizabeth Q., is yet another difficult patient—difficult not so much from a medical standpoint but for her behavior. My visit today provides a terrifying glimpse into how HIV can easily be spread even nowadays, more than fifteen years into the epidemic.

Although recently diagnosed with carcinoma of the cervix, a potentially serious gynecologic cancer, Elizabeth's ongoing drug addiction and self-destructive personality and her battering boyfriend are actually more life-threatening problems for her than her advancing HIV disease (her T-cell count is 12). Having been brought up-to-date by 3A's psychiatrist, I know a little bit about her tragic life of drugs, abuse, and abandonment, but I still cannot begin to fathom the pain she has had to endure. Sexually molested at age four by her father, Elizabeth was already using drugs by her early teens, when she began stealing and hustling on the streets to pay for her habit. Her present boyfriend, who is also her pimp, beats her whenever she does not bring in the money he needs to support his own drug addiction, and this present admission was the result of an especially bad thrashing that broke her jaw and collapsed her right lung. Of course, Elizabeth refused to file charges against her boyfriend, explaining that "he's on parole and will be sent back [to prison] if I call the police, and besides, he's all I've got."

A quiet, shy young woman who seems more like a helpless child than a hard-core crack user, she was transferred to 3A the previous week from the upstairs Spellman unit because she had stolen the cellular telephone of one of the nurses there. Never one for subtlety, Elizabeth was caught soon thereafter using the phone in her room to arrange a drug delivery. The nurse agreed not to press charges if Elizabeth was moved to another floor, and 3A was the lucky recipient of another unit's largesse. Once on 3A, she began to steal from her roommate—everything from food to clothes to money—and after several near-violent arguments with her roommate, who herself was an active drug user, Elizabeth had to be put in a private room, for her own protection as much as anything else.

Recently, Elizabeth's nurses have reported that she has been leaving the floor a couple times a day, usually for only fifteen minutes or so. I assumed she was only picking up drugs, but the staff can't baby-sit the patients and I did not make a big deal of it. During all of her adult life Elizabeth has turned to drugs to get her through her problems, and, I reason, it is way too late to expect her to change her ways and deal with AIDS *and* cervical cancer in a more stoical, socially acceptable manner.

But today, as my team and I are about to enter Elizabeth's room, one of the PAs reports that when he went in to draw blood from her earlier this morning, he caught her opening up the sharps container, the small plastic container attached to the wall of every patient room, where countless dirty needles from medication injections are disposed. These sharps containers, which are important safeguards in preventing needle-stick injuries to hospital staff, are usually locked and tamperproof, but the ingenuity of a desperate crack user can easily surmount such security precautions. Because she has never shot up with drugs before—crack cocaine has always been Elizabeth's drug of choice—I am almost afraid to ask her what she was doing in the sharps container. When I confront her, Elizabeth contritely admits that she has regularly been breaking into the container to get the tainted syringes and needles, and that during her brief elopements from the unit, she has been on the streets, exchanging these HIV-contaminated needles for vials of crack. Elizabeth Q.'s actions are perhaps some of the worst I have encountered on 3A. I remember an article in the Sunday *Times* about how the archdiocese and other guardians of public morality were heavily lobbying the city to discontinue the free needle-exchange program in the St. Clare's neighborhood, and I wonder how many people have been infected from Elizabeth's HIV-contaminated needles.

Elizabeth probably does not even understand that what she was doing was wrong—she says she "gets real nervous" if she does not have crack. Knowing that showing anger or disgust will accomplish nothing, and realizing that she still needs the protection of hospitalization, I strike a deal with her. I offer to prescribe her Xanax, a strong tranquilizer with high street value, if she swears to leave the sharps container alone and stop using crack. She readily accepts the deal and solemnly pledges to take Xanax instead of crack, probably thinking she is fooling me—but I am not fooled. I know she will, in squirrel-like fashion, hide away the Xanax and sell it on the streets for her precious cocaine, but I figure it will be a lesser evil than dirty needles. Such are the finer moral distinctions of practicing AIDS medicine on 3A.

After leaving Elizabeth, I briefly interrupt rounds to ask the head nurse to have Elizabeth's sharps container emptied as often as possible.

From Elizabeth Q.'s room, my PAs and I, like a portable medic team, move on to Harold M., who, like Elizabeth, has severe personality problems. A hulking, intimidating guy who appears healthier than his actual medical condition might indicate, Harold was refusing, as is his custom, to allow another important treatment plan I had arranged for him: surgical removal of his infected AV shunt. Diagnosed four months ago with kidney failure secondary to both HIV and active IV heroin use, Harold has been on outpatient dialysis three times a week. Characteristically, however, he did not show up for dialysis every so often, citing pressing "personal business"—that is, drug business. Because he had been recently hiding from the police after beating his wife, he missed his dialysis for over a week and consequently retained so much fluid that he developed congestive heart failure, which necessitated admission to Spellman two weeks ago for observation while his dialysis was restarted.

The previous week, a nurse found him in his bathroom, syringe and needle in hand, shooting up something into his AV shunt, the artery-to-vein connection that had been surgically inserted into his forearm to allow dialysis. Such a vascular graft makes IV injection with illicit drugs incredibly easy, which is a boon to someone like Harold, who long ago had used up all of his veins from active drug use. Sadly, this was not the first time Harold had done this to himself. Several months ago he had likewise damaged his original AV graft from shooting drugs into it, and that graft had had to be removed and replaced because of infection. Not surprisingly, as a result of last week's shooting escapade, the present graft quickly became red, tender, and pussy, and it needed to be removed before the infection spread to his entire body and killed him.

After several frustrating, time-consuming telephone calls—at St. Clare's, surgery specialists are notoriously balky about seeing Spellman patients—I had finally convinced a surgeon to see Harold and to agree once again to repair the infected graft.

"I want to wait till tomorrow, 'cause I've someone visiting me this afternoon," Harold says, not heeding my warnings that the surgeon would most likely walk away from his case if the surgery was canceled today.

Losing all patience and feeling tired of always having to plead with patients like Harold, I opt for a type of reverse psychology. I aloofly tell Harold that I really, in my own words, "don't give a fart" whether he has surgery or not—that he needs the surgery urgently, that the surgeon will drop his case if he refuses surgery on such a whim, that my conscience will be clear if he does not get his operation, and that if he dies from not having surgery, it is *his* problem, not mine.

"Harold, do whatever you want—it's *your* life, and frankly, I don't care anymore." Taken aback by my candor—after all, it is a bit frightening when the doctor appears not to care whether the patient lives or dies—Harold quickly backpedals, lamely insisting, "I never said I wouldn't have it, just that I didn't feel like it."

His bluff called, he says he will go to surgery today.

My beeper announces an outside telephone call. I am tempted to ignore the call and continue on with rounds, but because there is no telling how important, or unimportant, the call might be, I break off rounds and pick up the call. I am glad I did.

It is the mother of John R., who by coincidence is the next patient on rounds. Until now, I have been unable to talk with any of John's family, since the prison he came from did not have any phone numbers of family and thus could not notify anyone about his transfer to St. Clare's two weeks ago. Dogged detective work by Abby A. enabled us to locate his mother's address in rural South Carolina, and because there was no listed telephone number, a telegram was sent out last week, advising the mother about her son's admission and giving her my name and number to call for further information. Since John's condition is grave, I am glad to have a chance to talk with someone from his family.

John is a thirty-one-year-old African-American prisoner who for the past year has slowly been deteriorating from disseminated MAI, a common AIDS-related bacterial infection that infests almost every organ and results in a slow death from inexorable weight loss, fevers, and diarrhea. Although sick for many months in the prison, John had always refused hospitalization in the past, until he became so weak and unresponsive from his illness that he was finally unable to refuse transfer to St. Clare's any longer. Indeed, state prisons routinely ship out to the hospital such cases whenever it appears that death is imminent. Because the Spellman prison unit was full, John was admitted to Unit 3A, where two burly New York State prison guards have been standing watch outside his room twenty-four hours a day, despite the fact John cannot even hold a spoon or get out of bed by himself, let alone effect a daring escape from the hospital. As with so many of Spellman's prisoners with AIDS, John is now a prisoner of his disease.

Calling from a neighbor's telephone—the poor connection sounds as if it is from a distant country rather than backwoods South Carolina—John's mother wants to know about her son, saying that the telegram was the first word she had received about his even being sick, let alone being in the hospital. Before telling her any information, I need to know how much, or how little, she knows about her son's medical condition. With an outpouring of sadness I have heard so many times in the voices of other loving mothers, Mrs. R. relates how she lost touch with her son when he entered prison several years ago.

"He never wrote me back, and we don't have a phone out here for him to call us from the prison. I'm too sick myself to travel up North to see him," she explains almost apologetically, as if the lack of contact is her fault. It is quickly apparent that John's mother is a compassionate lady who is poor and who has no idea that her son has AIDS, let alone is dying. With a Southern deference that is touching, she politely insists I tell her the "whole truth" about her son's condition.

"The Lord gives me the strength to bear all things in this world, Doctor... so please tell me how my Johnny is." The pleading urgency, the self-confident strength, in her voice fill me with not inconsiderable awe.

I must instantly decide how much promise—and how much grief—to dole out to this lady. Do I tell her the entire truth—that her son is severely malnourished, terminally demented, and beyond all hope of even being stabilized from his downward spiral? Or do I speak in less painful generalities, varnish the truth, and spare her the pain, until he dies in a few days or weeks? Despite her plea for the "whole truth," I know, too, that telling her all the macabre details would be senselessly cruel, especially since she will probably never see her son before he dies. Moreover, despite her claims she can accept the "whole truth," I do not know her well enough to be certain how she might react—she apparently does not even know her son has AIDS.

Thus, I opt to tell her parts of the truth, but the crucial parts nevertheless: that her son has "full-blown AIDS," that he is "very, very sick and very, very weak" from complications of AIDS, and that, even though he is getting "the best care we can give him," there is no guarantee he will get better—rather, he will "probably get worse."

I spare her, for now at least, his pathetic prognosis and graphic description of his skeletal appearance. I do not inform her about the loss of his mind from HIV brain infection, the malodorous bedsores eroding his buttocks, his stool incontinence, and the intermittent sounds of moaning and groaning from nonlocalized pain and discomfort, all of which is requiring increasing doses of morphine. These aspects of the "whole truth" can wait until later, if she asks about them, or if she actually visits John before he expires. I figure that by now most people, even in rural South Carolina, know "full-blown AIDS" is not a sanguine diagnosis, and when she responds to my report with quiet, serious acknowledgments of "I see . . . yes, I understand," the gentle resignation in her voice tells me that she senses that her son is dying.

She again apologizes that both marginal health and limited finances prohibit her traveling to New York and asks if it is okay to call "from time to time" for updates on his condition. I reassure her that I understand her difficulty in coming to the city and that she should not hesitate to call anytime.

"And please tell Johnny," she concludes, "please tell him I called and that I love him, and tell him all the people in the church here will be praying for him round the clock . . . *please tell him that, Doctor.*"

Her voice conveys great caring, but none of the anxiety many AIDS mothers have about their sick daughters and sons. And her last words—that her church congregation in her small Southern town will be united in prayer—these words are spoken with a confident, matter-of-fact assurance that God will help her son. It is a serene assurance that their prayers for his soul are necessary and sufficient, regardless of whether God grants their petitions.

After finishing our conversation, I feel greatly moved by what I have just heard. This elderly lady, despite the barriers of time and distance, despite her son's past problems and neglect, still very much loves and cares about him. All she can do for him now is pray, and I believe that her love and prayers are probably as important as the gallons of largely ineffective antibiotics pouring right now into John's weary veins. I feel that the delivery of her message—that she loves him and is praying for him—is the most important thing I will do today, even though he may no longer comprehend the message. I am glad John is next on rounds.

Herding up the PAs to resume rounds—while I was on the telephone with John's mother, they were busy with the unit's endless scut work of drawing blood and starting IVs—I lead my team in to visit John. After quickly and perfunctorily examining him—as with many of my patients, he is so far gone that any such medical evaluation is purely pro forma—I bend over close to his ear and eagerly deliver his mother's message to him. He opens his eyes as I repeat her message several times to him, and—perhaps I am imagining it—he seems to comprehend what I have said, but without showing any emotion, any elation. I repeat his mother's message one final time, and he nods his head slightly and, too weak to speak, whispers vacantly, "Yes, my mother . . . my mother . . . mother . . ." His raspy whisper trails off without emotion, and his blank stare remains affixed to the ceiling. John's HIV dementia is profound, and what little cognition is left has been paralyzed by depression, fear, and the morphine that is dripping into his system.

As we exit John's room, his PA—an affable guy who means no harm—offhandedly remarks that earlier today one of John's guards divulged to him the crime of eight years ago that earned John a life sentence in an upstate maximum-security prison. As the PA is relaying the story to me and the other PA, my mind is largely preoccupied with thoughts about our next patient on rounds, and I do not immediately register what he is saying. Before I can protest to him that John's crime is none of anyone's business—in almost all cases, Spellman staff never know the past criminal records of their prisoners with AIDS—the PA blurts it out: "Word is that he sodomized a one-year-old girl. The guard said it was so bad she needed years of surgery to recover."

There is a silence that seems like an eternity.

I am momentarily stunned by this horrific news. My first impulse is to pounce with self-righteous fury on the hapless bearer of such a terrible message. But John's PA is a friend and did not toss off the news of the crime with relish, so I restrain my urge to upbraid him about how such confidential information is not to be bandied about. No, I realize my revulsion and rage are really toward John, not his PA, and for several minutes after being thus confronted, I must grapple with intense, conflicting emotions. Trying not to lose my composure, I numbly go on to the next patient on rounds, Joy D., who is recovering so well from PCP that she has been smoking in bed, while still on oxygen. Yet as I go through the motions of visiting her and absentmindedly admonishing her not to smoke, my mind is really still on John, his heinous crime, and the emotional turmoil that is churning my stomach.

Leaving Joy, I continue to feel confused. My thoughts even wander to my little nieces, in an involuntary attempt to personalize John's crime even further. To my horror, I feel a sense of righteous satisfaction, as if John is now reaping a punishment that matches the horror of his crime. These feelings of retribution fill me with disgust. Just fifteen minutes earlier, I hoped to assuage his fear and loneliness with news from his mother, and now I betray my deepest principles with thoughts of how he deserves to die alone and forgotten. Suddenly my concern and compassion for John are no longer abstract ideals, but rather are being tested by an infernal reality.

I have learned from previous emotional traumas to focus on the real purpose of my job—that the work I do on Spellman is important, not only for the patients and me, but also for a greater philosophic good. Namely, I am convinced that the medical work at St. Clare's validates the worth of *all* human lives, however marginalized and "worthless" society's indifference might render certain people. I believe that if the life of someone such as John is important beyond measure—and thus worthy of love and comforting—then *my* life and the lives of all human beings are of incalculable value. As trite as such sentiments might sound in the abstract, these beliefs are nowhere better illustrated, or more severely tested, than on a Spellman AIDS ward.

The turmoil now seizing my heart about John—that I both loathe him and care deeply about him—challenges my core beliefs. Although I know I will eventually be able to reconcile this conflict, I need more than just a few minutes to do so, especially since rounds are still unfinished. Too painfully cutting in my mind's eye are images of John's emaciated body lying in his deathbed, his loving mother praying with her church friends, and the brutally sodomized little girl. I need more time to reconcile these starkly contrasting images.

Resolving to revisit this turmoil later, I consciously put my emotions on hold and proceed on.

Next on rounds is Joey S., who, by sheer coincidence, is in his final death rattle as we enter his room.

Mercifully unconscious and pathetically wasted away to a skeletal figure of seventy-one pounds, Joey is slowly gasping the short, shallow breaths euphemistically termed agonal respirations. More like feeble sighs or shrugs, the involuntary upward spasms of his bony shoulders, clavicles, and neck are pitiful excuses for normal breaths. Joey is indeed shrugging off this world of suffering, sighing away his terrible afflictions from AIDS.

I now remember Joey's nurse telling me earlier this morning about the increase in his rectal bleeding. Joey's shallow gasps for breath just as we go in to see him do not alarm me. I know this will be an easy death for Joey, without all the wild and frenetic activity that attended Hector's departure an hour or so earlier.

In contrast to so many other dying Spellman patients, Joey's small room is crammed with family—his mother, three sisters, four brothers, and assorted in-laws. A thirty-two-year-old, third-generation Italian American from the Bensonhurst area of Brooklyn, Joey became a New York City cop after graduating from a local community college, and like his father and four older brothers—all of them city cops—he received several commendations for bravery in the line of duty. Several fellow cops and fortunate citizens would not be alive today if it had not been for Joey S. After his father was killed four years ago, Joey left the police force to study for the priesthood, which had always been an unrealized dream. However, when his superiors at the seminary learned of his pastoral work at Dignity—the city's gay Catholic congregation—he had to admit his homosexuality to them and was forced to leave the seminary. Soon thereafter, Joey, who had never been HIV tested, came down with his AIDS-defining illness, CMV retinitis, which rapidly progressed to total blindness. Having no health insurance and being unable to work, he had to go on welfare and live with his mother, who, along with her seven other children, gave him all the care and love possible.

Admitted to 3A only two weeks ago for bloody diarrhea and severe malnutrition, Joey was already too far gone to tolerate any invasive tests to evaluate his diarrhea—colonoscopy would probably have killed him, so fragile was his overall condition. Because routine stool tests were otherwise negative, I assumed that the bloody diarrhea was due to spread of CMV infection to his gut, an often fatal complication in patients as end-stage in their HIV disease as Joey. Unlike most Spellman patients, Joey arrived on 3A with signed Health Care Proxy, Living Will, and Do Not Resuscitate forms in hand—or, rather, in his mother's hands. He had long ago decided that when this moment finally arrived, he wanted comfort measures only.

I never really got to know Joey well, since his weakness was so severe that talking, even in a faint whisper, was too strenuous for him. But in my attempts to be sure the family was ready for the worst—which they were—I got to know his mother and sisters fairly well. Salt-of-the-earth Brooklynites, they provided Joey the loving support one would expect, but they would also buck up each other by recounting the happier family times they had had with Joey. Their reminiscences ranged from childhood outings at the beach to the seemingly countless christenings, first Communions, and weddings an Italian-American family of eight children and innumerable grandchildren would have. And all of their stories—whether about a school play from many years back or a vacation to the south Jersey shore—would contain a particular remembrance, a fond anecdote, about Joey and how his presence made every family event a little happier, a little more special. Joey had always been "the baby" of the family, and his secret sexuality was not discussed and was never an issue: he was family, and that was all that mattered.

Although the death will be relatively easy and painless for Joey, it will not be a pretty one for everyone else at his bedside. Indeed, most deaths I have seen on my AIDS ward have probably been nonevents from the patients' perspectives. Rarely is there dramatic agony, with the patient's face contorted in discomfort, or with cries of excruciating pain. Rather, the fearful expectations of death from AIDS—or from anything else for that matter—far exceed the actual event. The final joke in life is undoubtedly the *ease* with which we will die, much along the lines of the old song "Is That All There Is?"

The problem most people have with death, of course, emanates from fear—fear of pain, fear of incapacity, even fear of fear itself. Much of this fear comes from having watched other people die and assuming—falsely—that the dying feel as bad as they appear. While pseudoscience makes a great deal out of so-called near-death experiences, I do believe that the body can experience an intense sensation of pleasure or completion (others might call it a religious feeling) at the point of death. Our species has actually evolved remarkable protective chemicals called endorphins, which the brain produces in times of stress and pain, to relieve the challenges of dying. Of course, these endorphins mitigate only the dying person's perception of death, while the rest of the body can often appear dreadful, much to the consternation of those still living. As he lies dying before his family, Joey does not feel as bad as he looks.

For the past several days, Joey's increasing rectal bleeding has necessitated almost continuous blood transfusions, primarily because his family and I felt that transfusions might keep him more comfortable. As his nurse reported early this morning, the bleeding is now unending, an uncontrollable torrent, which, just as my team and I enter the room, is starting to stain through his sheets, overflow his bed, and steadily drip onto the floor and around the edges of the bed. Like an almost unimaginable scene out of a horror movie, the sanguinary flow steadily increases before everyone's eyes, slowly dribbling into several bloody puddles on the floor. It is as if the two units of blood being simultaneously pumped into the veins of his arms are being directly siphoned into the bed and onto the floor.

My team and I stand just inside the doorway, transfixed by the sight. Besides the blood, small shreds of granular, tissuelike material cling to the metal bed frame. The scene defies easy comprehension—is it merely stool or can it even be part of the inner lining of his colon, I wonder?

Neither I nor my PAs have ever witnessed such a grotesque spectacle. Jolted from our morbid trance, we nervously try to negotiate around the family members gathered at the bedside as we attempt to cover the bloody puddles with towels and blankets. Our efforts produce little effect: the steady flow continues unabated. Whereas we are inexplicably embarrassed by the trauma, Joey's family appears hardly to notice this final, massive outflow of blood. Oblivious to the blood dripping onto their shoes, his mother and sisters are standing close at bedside, holding his pale, skeletal hands, which clutch several sets of rosary beads. Standing close behind is a retinue of brothers and in-laws, and the family priest stands off in a corner quietly reciting the Office of the Dead. The musky odor of fresh blood overwhelms the unair-conditioned room, which is cruelly illuminated by the bright August sun burning through the dirty window shades.

Joey's jerking breaths now grow scanter, and pathetic little gurgling noises from his nose and mouth can barely be heard over the quiet sobbing and sniffing of his family. Standing on his bedside table is a picture of him and his parents the day he graduated from the police academy. Both he and his father are in uniform, and draped over the picture frame are the multicolored medals of valor he received as a city cop. Taped on all the walls of the room are numerous get-well cards and—saddest of all—a multitude of get-well pictures and drawings from his countless little nephews and nieces: "Please get well soon, Uncle Joey," "We love you, Uncle Joey," and so forth, all conceived in the boundless spirit of hope that only a child can have in the face of AIDS. Outside, on the sidewalk three stories below, a passing group of schoolchildren cackle and laugh on lunch break, and Joey sniffs in his last breath and dies.

His family knows. They do not need me to tell them. Soft sobs waft throughout the room. The blood will of course trickle on for a while, but mercifully, the two units of blood being transfused are almost empty, obviating the need to dramatize his death by turning off the transfusion. Approaching the bedside, I choke out some stock condolences to his mother and sisters. All of us know there is nothing I, or anyone else, can say to comfort them. Haggard but still unbowed, his mother tightly holds and shakes my hands. She profusely thanks me for Joey's care, her daughters joining in with their gratitude. I feel embarrassed but try not to show it. I know that Joey's AIDS was too far advanced for me to have done much of anything for him.

Leaving for the nurse and PAs the all-too-familiar routine of death certificates, body bags, and funeral home arrangements, I exit Joey S.'s room for the last time and go on with my rounds.

The next patient on rounds, Mr. Ralph P., presents a classic Spellman dilemma—not a story of death and dying like Joey S.'s, but an all-too-common example of the difficult and frustrating choices I must make daily on my AIDS ward.

Forty-two years old and homeless, Ralph was admitted to 3A several weeks ago for a fairly routine bacterial pneumonia that promptly responded to antibiotic therapy. But, as with so many of my patients, Ralph's case quickly became more complicated as soon as his immediate, life-threatening problem resolved. On his physical exam at the time of admission, I noted a few purplish skin spots on his ankles. He said they had been increasing in size and were becoming more painful, to the point he sometimes had difficulty walking distances. I promptly asked the part-time Spellman dermatologist to see Ralph and biopsy one of these purple spots, the assumption being that they were probably Kaposi's sarcoma, the AIDS-related cancer, which might perhaps respond to radiation therapy. Like several of Spellman's consultants, the dermatologist, engaged in better-paying or less unpleasant work elsewhere, often has to be asked more than once to see a patient, and after the second proding—a full week later—she finally saw Ralph and biopsied the ankle spots. But the delay in making a diagnosis for Ralph's skin lesions was not over yet. Because St. Clare's pathologist felt "uncomfortable" making a definitive diagnosis from the biopsy specimen, it was forwarded to an outside dermatopathologist for a reading, which, according to the lab, would not be back for another four or five days.

In the meanwhile, as these delays in diagnosis were piling up, Abby A. was trying to locate some decent housing for Ralph, so he would not be tempted to return to the streets after being discharged. Sadly, Ralph is one of the legions of New York's homeless who—probably correctly—feel safer on the streets than in one of the city's homeless shelters, which are often plagued by drugs and violence. As a result of Abby's efforts, he has been accepted as a resident at Gift of God, a highly sought-after AIDS home in the city. Earlier today, Abby reported that Ralph is scheduled to be transferred to Gift of God tomorrow. If he cannot go tomorrow as scheduled, the bed reserved for him will quickly be snatched up by some other homeless AIDS patient on the waiting list, and Ralph would then have to be discharged to a homeless shelter while the city's understaffed Department of AIDS Services tried to find more permanent housing for him. This latter possibility—being discharged to a shelter—would almost certainly mean Ralph would end back out on the streets, where he previously neglected himself and developed pneumonia.

Although I am happy to hear that Gift of God will take Ralph—he is an affable guy who just needs minimal supervision to stay off the streets and away from drugs—I also realize that, if he does go to Gift of God tomorrow as scheduled, it is highly unlikely there will be any further evaluation or treatment of his painful ankle spots. Forwarding information such as his pending biopsy result from St. Clare's to an AIDS home such as Gift of God is usually impossible, primarily because of bureaucratic red tape and rigid state laws prohibiting transfer of HIV-related diagnoses from one health-care facility to another. As incredible as it sounds, if Ralph goes tomorrow, his doctors at Gift of God, who generally see patients there only once a month, would have to repeat the skin biopsy, resulting in even more delay in diagnosing and treating his increasingly painful ankle spots.

Thus, the dilemma for Ralph and me is frustratingly simple: Do I delay his discharge tomorrow to wait for the biopsy report to return, so that the cause of his ankle pain can be properly diagnosed and treated, but at the risk of his losing his bed at Gift of God and ultimately ending back out on the streets, or, do I opt for discharging him tomorrow to Gift of God, but at the expense of postponing evaluation of his probable KS, which could well worsen during this delay?

What I would really like to do is keep Ralph in the hospital until the biopsy returns and appropriate treatment is begun, and, once that is done, continue to keep him on the ward until another bed at Gift of God (or a similar facility) opens up. Indeed, since he is quite content just to "hang out" on 3A, such an option is just fine with him. However, the hospital "utilization review" staff are already giving Ralph's case the so-called hairy eyeball and are pressuring me to discharge him as soon as possible, thus making my optimal game plan untenable. I can hold off the utilization-review bureaucrats until the biopsy returns and necessary treatment is begun, but not after that. On rounds today I try to explain to Ralph the quandary facing both of us, but the subtleties of the choices are too confusing for him.

"Do whatever you think would be best for me, Doc," Ralph blithely replies.

The rank absurdity, the patent unfairness, of Ralph's dilemma angers me, but I try not to show it as I discuss it with him. It would only make him uneasy and more confused. In a *real* hospital, I fume to myself, Ralph's problem would have been dealt with *weeks* ago. In a *real* hospital, the dermatologist would have seen and biopsied the patient within a day, the pathologist would have felt "comfortable" reading the specimen the next day, and the following day radiation therapy to the biopsy-proven KS spots would have commenced. By this point in his hospital stay, Ralph would have been in much less pain . . . in a *real* hospital, that is. Instead, Ralph is screwed by the incredible inertia of "the system" and its bureaucracy.

But my anger today is tempered by both fatigue and the realization that Ralph's case is only a minor atrocity. After all, I think to myself, "only" prolongation of a patient's KS pain versus consignment to the streets is involved, and not something immediately life-threatening. Too perturbed and too tired to choose the lesser of two evils for Ralph, I decide to postpone any final decision until tomorrow, which at this point in the day seems like light-years away.

As I plod along on the rest of my rounds today, most of the remaining patients are not as dire as the previous ones. For whatever reason, most of the problems of titanic proportion usually blossom early in the day, allowing the final hours to be spent in more of a mop-up mode. Moreover, on the remainder of rounds today there are even some success stories, at least by St. Clare's standards.

A few hours ago, Orlando V.'s sputum smears finally returned negative for active TB, and he can be released from respiratory isolation, although his brief escape hours earlier to a corner deli for breakfast makes the issue academic at best. Hopefully, his compliance with his TB medicines will be more exemplary.

Antonio L.'s eyesight has stabilized now that he is receiving IV ganciclovir for his CMV retinitis. In fact, his vision has improved to the point that he can resume his hobby of watercolor painting—on rounds this afternoon, he says he is going to start painting a picture of me. As I conclude my visit with Antonio and feel satisfaction about his progress, I quietly wonder, and worry, whether his CMV will remain in check or will progress further, as happened with Joey S., whose discreetly covered body is on a gurney and is just now being rolled onto the elevator on its way to the morgue.

Narvor H. is now resting comfortably on a low-dose IV morphine drip for the brain tumor and meningitis he is slowly dying from, too obtunded to any longer disconnect his IV antibiotics. On his bedside table is a note left earlier in the afternoon by Carmen, his younger sister: "I love you, Narvor. . . . I was here and kissed you. . . . I'm sorry I argued with you in the past." A few months ago, before being stricken by these final complications of AIDS, Narvor had just been released from prison, and despite having been estranged from his family for many years, he was taken in off the streets by this sister. Carmen was the only one in her family willing to give him another chance, and she let him live with her in her small Bronx apartment. However, when Carmen left shortly thereafter for a brief visit to their parents in Puerto Rico, Narvor shamelessly stripped the apartment bare—down to and including lightbulbs and new copper wiring and pipes—to finance a drug binge, which ended when he was brought to St. Clare's for intractable seizures. Carmen's understandable rage did not cool until I called her late last week about the utter despair of her brother's condition, and all was again forgiven. Up until my call to her, she had not visited him on 3A. Unfortunately, Narvor, who already suffered fairly advanced dementia, has been increasingly confused and lethargic for almost a week, essentially unable to understand his sister's attempts at reconciliation.

By midafternoon, the two patients who had been demanding narcotics had been calmed down. The young woman who had been loudly demanding an increase in her methadone as soon as I arrived this morning is now placidly nodding off on her extra ten milligrams of drug, and when I confront the barefoot heckler who was hassling his nurse for more Percocet, he indignantly denies any rudeness on his part.

And finally, Roberto Q. admits to "sometimes" squirreling away his TB pills, for reasons he is at a loss to explain. Despite my weary, and by now stereotyped, attempts to encourage him to take all of his medicines, he is still much more intent on impressing upon me how aggrieved he feels about the mix-up with his breakfast order earlier today.

"I gotta have eggs, Doc, 'cause oatmeal gives me bad gas," he whines.

I dumbly look at him as he obsessively prattles on about his intestinal gas and his diet. Somehow, at this point in the day, it makes perfect sense that Roberto's breakfast tastes are far more important than his TB treatment.

Patient rounds at last completed, I now set out writing the daily progress notes in the patients' charts. Indeed, although St. Clare's is sometimes backward in many other aspects of "modern medicine," the legalistic obsession for chart documentation and mindless paperwork is one area where the hospital is state-of-the-art. The hospital's various "quality assurance" committees, as well as government regulators and hospital accrediting agencies, judge the quality of medical care at St. Clare's by the progress notes and other paperwork in the patients' charts, not by the actual time and efforts spent with the patients. "Quality of care" in the mid-1990s devolves entirely on the written word, even if it bears little resemblance to reality—the stock motto of quality-assurance bureaucrats is, "If it [medical care] is not documented [written in the chart], then it didn't happen."

Because I deal daily with the frustrating imperfections of medical care at St. Clare's—the filthy rooms, the overworked nurses, the shortage of simple supplies such as slippers and toothpaste for the patients, to name only a few privations—I sometimes chafe under the hypocrisy of the hospital's spending precious resources on so many "quality-assurance" activities, since *real* quality care suffers at the expense of maintaining the *appearance* of quality care. But, as with most of my piques, this trifling annoyance, which is not unique to St. Clare's, saps my energies and just distracts me from the work at hand.

Late in the afternoon, just as I have finished all my charts and other paperwork, an ambulance crew hurriedly wheels Ricardo C. off the elevator and back onto 3A to be readmitted, as I advised his mother earlier this morning. Appearing to be asleep, Ricardo is enshrouded by the covers on the ambulance stretcher, and were it not for his gaunt head poking out at the top of the stretcher, one would not even know there is a thin human body hidden under the folds of sheets. As always, his mother is at his side, her eyes weary with concern. She tries to keep up with the ambulance attendants—ambulance crews are *always* impatient and in a hurry—who are briskly pushing the stretcher down the hallway, ironically to the same bed he was discharged from only a week earlier. A tall, almost domineering lady with a seriousness of purpose in her stride, she is carrying in one hand a small suitcase, and under her other arm Ricardo's two large stuffed animals—a pink rabbit with copiously floppy ears and an oversize panda—that were with him during the last admission. Ricardo previously worked as a movie technician in Hollywood, and the stuffed animals had been a get-well gift from his coworkers in California. I allow the nurses a few minutes to transfer him to his bed and to check his vital signs, before going in to see him. I already know that the real challenge will not be Ricardo and his end-stage AIDS.

Indeed, no sooner do I enter Ricardo's room than Mrs. C. immediately accosts me, seeking urgent reassurance that I cannot give.

"He's going to be all right, isn't he, Doctor?" she inquires—and insists—in her typical manner that is simultaneously a demand and a plea. It is as if my saying Ricardo will be okay is all that matters.

I do not reply but smile my standard smile of earnest concern, approaching the bed to examine him. Ricardo is unconscious and unable to give any history. Pulling back the bedcovers to begin my exam, I behold his skeletal frame covered by barely adhering. bloodless-appearing skin—no muscle or fat is left on Ricardo's body. He looks much worse than when I last saw him a week ago. How many different words, I muse—how many synonyms—have I had to conjure up over the past years to adequately describe the pitiable appearance of AIDS patients like Ricardo? *Profoundly malnourished, terminal, profoundly wasted, skeletal, end-stage, cachectic, emaciated,* have been adjectives I have used time and time again in my hospital notes to describe such patients, but after a while mere words lose their meaning and impact.

Ricardo is curled up in the fetal position, with his head bowed down to his chest, his arms and hands folded tight up against his chest, and his legs bent in the knee-to-chest position. Practically every joint is frozen in place from stiff flexion contractures, resulting from many months of inactivity, as well as malnutrition. Several large stinking bedsores on his buttocks have eroded through his fragile skin down to the bone and ooze pus that is mixed with dried blood, maggots, and stool. Ricardo's entire body is filthy and malodorous, with stool smeared even in his thinning, straggly black hair. The attempt at "home care" has been an unmitigated disaster.

The remainder of my physical exam is cursory and perfunctory. I go through the motions of listening to his heart and lungs and palpating his practically nonexistent abdomen all the way to his backbone, primarily for the benefit of his ever-watchful mother—to make her feel that *something* is being done for her son, even though I know medical science can do absolutely nothing to save him. Indeed, I think to myself, if Ricardo C., in his present condition, were suddenly deemed to be the most important human being on earth, and if he were somehow magically surrounded by the world's finest medical specialists, nurses, and hospital care—if he were hypothetically attended by such a stellar staff around the clock and closely monitored and provided the most intense and state-of-the-art medical care, he would still rapidly succumb to AIDS, and all of this urgent and flawless care would still be for naught. Thus, short of active euthanasia or intentionally inflicting pain, I or anyone else could not hurt Ricardo or make a mistake in his care. Regardless of what I do or do not do to Ricardo, the outcome will be the same. As with many of my AIDS patients, Ricardo C. presents me with more of an ethical problem than a medical one.

Intently watching my ministrations, his mother stands at the foot of the bed, still clutching the stuffed animals and the small suitcase she brought along to the hospital. As on prior admissions, Mrs. C. will undoubtedly be camping out in Ricardo's room twenty-four hours a day, since she has no other children or other close family to preoccupy her, her husband having died several years ago.

"He's going to be all right, isn't he, Doctor?" she anxiously presses me again, the demand and desperation in her voice sounding even more acute than before.

Sometimes, as in situations such as this, I feel my only job on 3A must be to take away forever whatever feeble hope patients and their families are clinging to. Which is more cruel, I ask myself just now—for what seems the millionth time on Unit 3A—giving Mrs. C. false hope about Ricardo, hope that will quickly be crushed when he dies in the next few days, or being brutally realistic from the very start?

Recalling quite well Mrs. C.'s previous refusals to accept Ricardo's terminal state, I decide to avoid any strident discussion of how hopeless his condition is, but I am not going to encourage her illusions. Mustering as much compassion as I can this late in the day, I look Mrs. C. directly in the eyes and softly but firmly reply, "No, Mrs. C., Ricardo is not going to be all right. He is very, very sick from AIDS and is not going to get better. We need to keep him comfortable and out of pain."

As with my prior conversations with her, Mrs. C. does not react to my bleak assessment, and she nervously starts to chatter on about problems she was having with Ricardo's home-care nurses, as if minor matters such as the frequency of dressing changes on his bedsores would have made any difference whatsoever in his medical condition at this stage. As in the past, Mrs. C. prefers to dwell on unimportant matters and does not want to hear the truth. I don't resume the struggle to prepare her for the loss of her only son, her once-handsome son who was so full of promise in Hollywood, and who has been the central focus of her life for the last year. My mind wanders to Hector R.'s atrocious CPR earlier today. I despair to think that the same farcical charade awaits poor Ricardo in the coming days. He, too, is not a DNR, thanks largely to his mother's insistence that "everything possible" be done. Her hope for a "miracle" is once again untempered by reality.

Finally excusing myself from Mrs. C.—she is one of those family members who could talk on forever—I return to the nursing station and commence Ricardo's admission paperwork. When I come to deciding what tests and treatment to order for him, I face yet another ethical dilemma. I have major qualms about wasting money on tests and treatments that will not materially prolong Ricardo's life, but I also realize that, as with many of my AIDS patients, I am not just treating Ricardo, but also his mother. She will expect the blood tests, the IV fluids, the X rays. I rationalize to myself that ordering these tests and treatments will probably not prolong his suffering, especially since I am also ordering liberal doses of morphine. Thus, as I often do on my AIDS ward, I again take the path of least resistance and order lab tests and treatments for Ricardo, rather than later face the barrage of questions from Mrs. C. as to why I am not doing "everything possible" to help her son.

• • •

At the end of the workday on Unit 3A I briefly review all the unit's patients with my two PAs.

As we go down the roster of patients, we try to make sure each one's major problems, lab tests, and X-ray results have been addressed—not necessarily solved or corrected, but at least addressed. When my team and I finally do walk off the floor for the day, we always leave behind what seems like a thousand loose ends: abnormal lab results, unfinished X rays, untreated patient symptoms, unresolved diagnostic dilemmas, and unassuaged loneliness and emotional pain.

Yet, before leaving, I always try to be sure that we have at least dealt as best we can with the major, life-threatening problems on 3A that day: the blood infection that can kill in a matter of hours, the meningitis that can quickly progress to coma and death, the severe pain of cancer, the worsening pneumonia that can overnight lead to terrifying breathlessness and respiratory failure. The multitude of less urgent problems—the blood work that could not be drawn because of either patient refusal or lack of veins, the needed X ray that was not done because either the patient refused or the X-ray department never got around to it, the patient who continues to refuse crucial medication for this or that opportunistic infection, the nursing-home application form that the social worker says *must* be done—all of these less urgent matters will have to wait until tomorrow, or perhaps even the day after. I could spend every waking—and sleeping—hour on 3A and still not make a significant dent in all the problems there. My operative phrase, my 3A mantra, is that my team and I do the best we can with what we have available, which is often not very much.

Indeed, although I still believe that medical science can sometimes make a difference, even in AIDS, I also feel that my colleagues and I really make a difference at Spellman just by being there. Our reassuring presence, our kind words to patient and family on daily rounds, our just showing up for work each day, often matter as much, or even more, than many of the medicines and tests we order. When the insanely hectic pace of 3A obscures this important fact for me, I have to stop and put my work there in proper perspective. Indeed, the only loose ends I really hate to leave unattended are the largely intangible ones, the kind that do not show up in chart notes or quality-assurance audits: the supportive words to patient or family that should have been spoken but have not been, the extra visit to a dying patient to be sure enough pain medicine is being given, the final good-bye to a patient I feel will not make it through the night. Sometimes, I leave 3A with these important loose ends taken care of; other times, because of the impossible press of time, regrettably not.

Today as I survey the ward one last time, I feel that I have tried to address the major emotional needs of my patients and their families. The search is on for Jose T.'s mother, and Jose senses that 3A and its staff are not the enemy. Hector R.'s mother has been given my best effort at emotional support; presumably, her self-blame about not being at his bedside when he died will quickly pass. The unalloyed message of love from John R.'s mother in South Carolina was faithfully delivered. Joey S.'s mother and sisters are probably dealing with their grief together. On and on, my mind replays the major events of the day as I finally walk off Unit 3A for the day.

Emerging from the hospital's dreariness into the frenetic street activity of the outside "real world," I feel an honest sense of exhilaration. The sidewalks of Hell's Kitchen this late-summer afternoon are teeming with a cross-section of New Yorkers: gossiping old men harmlessly drinking cheap beer on street corners, well-dressed young professionals hurrying from work to nearby ethnic restaurants for dinner before the theater, the homeless rummaging through trash containers next to Korean grocery stores, drug dealers exchanging money and heroin, and the obligatory panhandler. I marvel at the contrast between the death and decrepitude within St. Clare's and the vitality of human activity on the streets right outside the hospital.

This stark difference between what is inside and what is outside St. Clare's invariably invigorates me at the end of the day, dissipating any anger or despair the day's work may have caused. I could never leave St. Clare's with an exhausted feeling of relief or defeat any more than I could go to work each morning with a sense of foreboding dread. Rather, as clichéd as it sounds, I take away from work a profound appreciation of how special my life, my patients' lives, *all* human life is.

This sense of life's preciousness, which I inhale from the streets on my way home from work, is how I can survive the daily spectre of disease and death. It is how I can maintain my emotional perspective day in and day out. Not only do I never fear the daily reminders of my own frail mortality, but I am grateful for this exposure, this realization that *we are all ultimately HIV-positive,* in that we are all going to die sooner or later. I acutely realize that someday, regardless of my own final disease or injury, I, too, will join my many patients on their sickbeds. The poignant stories transpiring every day on my AIDS ward, my crucible of despair and hope, have taught me that living a life that denies the relevancy and imminency of death actually robs that life of the wonder it should really have. Indeed, I have come to believe that a content life is one that gracefully carries death on its shoulder as a friend and not a feared adversary.

By the time I arrive home today, I have largely put out of my mind the stresses of the day's ordeal on 3A. Rarely do I let the problems of work fester at home. Yet tonight, as I go about my routine, my mind drifts on and off to several searing images of the day: the new admission, Jose T., alone in his squalor . . . the week's first death, Hector R., and his disgusting, violent resuscitation attempt . . . the pathetic prisoner, John R., as well as his sweet mother in South Carolina and the young victim of his terrible crime of eight years ago . . . Joey S.'s bloody death, "horrendous" for the living but not for the person dying . . . and Ricardo C.'s impending death, with his skeletal body frozen in the fetal position and pitifully disappearing into the folds of the bedsheets . . .

Later in the evening, as I lie awake in bed, I think about the many mothers from earlier in the day. I wonder if Jose T.'s mother is still alive in Puerto Rico, and if so, will she even care that her long-lost son is sick with AIDS? I wonder how Hector R.'s mother is coping with at long last being home alone, no longer holding twenty-four-hour vigil at her son's bedside. I remember how forlorn she looked as she left 3A for the last time. I visualize in my mind's eye John R.'s mother in South Carolina, praying with her church friends for the son she has not seen for so many years. I see Joey S.'s dark red blood dripping onto his mother's shoes at bedside as she concentrates solely upon being at his side as he breathes his last. I wonder how she reacted when she later saw her son's dried blood on her shoes. And I think about Ricardo C.'s mother, who is there with him now and is undoubtedly puttering about his room, adjusting his bedcovers, propping up his toy animals, and anxiously reacting to his every moan.

A strange mixture of sadness and wonderment, of love and suffering, overwhelms me. Tired and ready for sleep, I roll over in bed and snuggle up with my little pet dog. I feel incredibly blessed, incredibly fortunate, to be an AIDS doctor.

3

A GRAND CONCOURSE

"Rosa M." . . .
"Yolanda J." . . .

Each name called from the altar, an entire universe of memories . . .

THIRTY-ONE-year-old Rosa M. reached her own fifteenth ring of hell the winter night police found her unconscious and half-naked in a darkened subway tunnel. She was clutching an empty crack vial, lying off to the side of the tracks in puddles of mud, excrement, and vomit. Later that same evening, she was admitted to St. Clare's AIDS service for the first and last time. It was a night so distant from more halcyon days twenty years removed, when she was a little girl in the South Bronx, and a night even further removed from the peace that would transfigure her five weeks later, immediately prior to her death.

Rosa's journey—from childhood walks with her grandma along the Grand Concourse Boulevard, to her subhuman existence in the subway catacombs decades later, to her rebirth right before her death—was both transcendent *and* mundane: transcendent in its miracle of lost love reclaimed, mundane in how Rosa M. epitomized the prototypical St. Clare's AIDS patient.

A month before her descent into New York's subterranean underworld, Rosa had been paroled from an upstate prison, where she had done one and a half years for drug possession. Although released to a Midtown halfway house for parolees, Rosa immediately returned to the city's innumerable crack houses, trading sex—mostly unsafe, "skin-to-skin" sex—for heroin and cocaine, just as she had done for many years before her imprisonment.

Rosa had first turned to drugs at age thirteen, not long after her mother and father had perished in an apartment fire, innocent victims of one of the many drug-related arsons that had plagued their South Bronx neighborhood. On the night of the fire, Rosa was staying a block away with her maternal grandmother, who was both her namesake and her closest friend. For as long as both of them could remember, Mama Rosa—Rosa's nickname for her grandma—had always doted on her granddaughter, taking her shopping, to the laundry, to church for mass and rosary readings, or, to little Rosa's irrepressible delight, on long walks up and down the Grand Concourse Boulevard, shabbily magnificent with its wide avenues and rows of fading art deco apartment buildings.

Two generations earlier, long before a two-year-old Rosa arrived from San Juan with her family, the Grand Concourse had been a vibrant urban thoroughfare, an open-air forum for the culturally diverse citizens of the South Bronx. The Grand Concourse was exactly that—a four-and-one-half-mile boulevard of monumental proportions, 182 feet wide, with two tree-lined dividers and broad sidewalks that fronted the borough's major shopping and entertainment district. At one time—before drugs and poverty decimated the community—desirable apartment buildings lined the avenue and comprised one of the largest art deco urban sprawls in the United States. Home to an aspiring blue-collar and middle-class Jewish population, the Grand Concourse had been the home of a cross-section of American celebrities, including Roberta Peters, Babe Ruth, and Milton Berle. The post–World War II era ushered in the present period of decline, and the previously Jewish residents were replaced with poorer African Americans and Hispanic Americans. Despite the seemingly inexorable decay from drugs and crime, the Grand Concourse still retained a vibrancy and magic in the imaginations of some of its newer denizens, including Rosa M.

In the weeks before her death at St. Clare's, Rosa M. would fondly recount to both Sister Pascal and Abby A. those childhood outings with Mama Rosa—how Grandma would always stop at the same corner bodega for her Rosa's favorite candy, how Grandma would proudly show her off to old ladies they would meet on the Concourse, and how safe, and indeed how *important,* Grandma always made her Rosa feel.

"The Grand Concourse was like magic to me," she once rhapsodized to Sister Pascal, herself a Bronx resident. "The best time was the summer, right before and right after sunset . . . the buildings just *glowed* in the light . . . there was pink, orange, even green. The Concourse seemed so big to me—it seemed to go on forever and ever . . . and all the people on it, coming and going . . . I loved all the movement, the excitement of all the different people there. And all along I'd have my Mama Rosa right there with me, telling me stories as we walked. They were always the same stories, but I loved hearing them over and over, stories about when *she* was a little girl in Puerto Rico and how she loved *her* grandma."

Rosa M.'s dependence on her Mama's love was the central, immutable underpinning of her childhood. Much later, when she was a Spellman patient and lay near death, her sister placed a faded family photograph by her bedside. It showed a plain, shy-appearing seven-year-old Rosa, dressed in a frilly pink dress and sitting on her Mama's lap, the two of them turned almost imperceptibly away from her parents and older sister. Little Rosa's expressive brown eyes suggested a forlorn appearance, while her smile looked tentative and fragile. A serene-appearing Mama Rosa dominated the photograph, and she proudly cradled her granddaughter in both arms, one of which little Rosa clung to. Like wax figures, her parents and older sister, Maria, appeared stiff and formal, sharply contrasting with Mama Rosa's steadfastness and Rosa's vulnerability.

Rosa recalled that the happiest day of her life was her first Communion at the age of eight. As a result of Mama Rosa's handiwork, Rosa's white dress was the prettiest one at church, and Mama Rosa seemed transported to paradise when her little Rosa walked up to the altar. To commemorate the day, Mama gave her a set of rosary beads and a new Bible, with the inscription, "I will wait for you in Heaven, my little Rosa. We will walk together with Jesus . . . Love, Mama Rosa."

"I never wanted to think about Mama ever dying," Rosa later explained to Sister Pascal, during one of Pascal's many visits in the weeks up to Rosa's death. "Once, when I was real little, I was lying in bed next to Mama late at night—Mama and I always slept together—and I heard sirens in the distance. They sounded so sad and lonely, and I thought to myself, 'Mama's gonna die someday.' I got all upset and started crying. Mama woke up and told me I had a nightmare. She gave me a hug and said, 'Your Mama's here, baby, you're with your Mama,' and after that I never thought about Mama ever dying."

Tragically, the very evening after her parents' funeral, Rosa M. found her Mama Rosa dead, slumped over in her sitting chair, rosary beads in hand. Everyone assumed the shock of her daughter's and son-in-law's deaths was too much for her heart to bear. Rosa M.'s grief was profound, overshadowing her conflicted emotions about her parents' demise.

"Mama Rosa was my *whole life,*" she later told Abby A., who, as Rosa's social worker, visited her frequently during her stay at St. Clare's. "I don't even remember much about her funeral, I was so hysterical for most of it. I didn't eat or go to school for weeks, until Maria made me go."

During the months after Mama's death, Rosa lived with Maria and her husband. Ten years older than Rosa, Maria was as cold and aloof as their parents. She did not indulge her little sister, and soon Rosa and Maria were fighting over curfew hours, schoolwork, and the delegation of household chores. Rosa was expected to baby-sit her two-year-old nephew, not to watch too much TV, and not to socialize with the neighborhood boys, who had recently caught her adolescent eye. Instead, Rosa started missing school and even stopped going to mass—just walking past the church evoked painful memories.

In the months after Mama's death, Rosa tried to blot out any thoughts of her paradise that lay irretrievably lost. When some neighborhood boys shared reefers with her during one of her melancholic moods, she liked the feeling—very much. The hurtful memories receded, and life became tolerable, even transiently pleasurable. Experimentation with crack cocaine and heroin—both more readily available on her block than fresh milk—proved more potent, enabling her to cope with the loneliness, at least for a while. Quickly, decisively, the drugs became the justification for Rosa's existence. The ongoing arguments with Maria escalated with Rosa's drug use. Drugs, Maria insisted, would *not* be tolerated in her house, and soon thereafter thirteen-year-old Rosa was living on the streets, more by mutual agreement than by forced eviction.

The subsequent sixteen years were a nightmarish blur. To support herself and her habit, Rosa quickly turned to the sex-for-drugs trade, rapidly cashing in her youthful appeal for the drugs that would briefly expunge her pains. Rosa's life was transformed into an endless procession of sadistic "boyfriends," who pimped and abused her, of short drug highs followed by hellish crash landings, of several botched suicide attempts, and of interminable years—over one and a half decades—trolling the city's crack houses for drugs and sex. Rosa's degradation as a prostitute knew no limits as she furtively serviced countless johns in back alleys, bar bathrooms, and by-the-half-hour motels. Her hands, arms, and legs became scarred over with thickened needle tracks. Addled by the drugs, she lost track of her abortions and frequent episodes of venereal disease. She would, she said, regularly have sex with a pharmacist in exchange for antibiotics to expunge her periodic pelvic pain and genital sores. Rosa's three major drug overdoses resulted in her only contacts with medical care, but the staff at Bronx Lebanon Hospital were unable to cope with her drug addiction since, invariably, she would stagger out of the hospital as soon as she recovered consciousness. Her gaunt face, sunken eyeballs, and dull gaze suggested a soul beyond redemption. The heavy makeup, spandex miniskirts, and florid eyeliner adorned a zombie—a walking corpse, not a young woman who only a few years earlier had escorted her Mama Rosa along the Grand Concourse to mass and evening rosary every day.

Prison completed the annihilation of Rosa M. Arrested for selling crack to an undercover cop—she needed money to pay off an impatient loan shark, who had already broken her nose—Rosa ended up in Greenhaven Prison, a minimum-security facility north of New York City. Before she was arrested, her need for drugs had rendered her more prey than predator, as her periodic black eyes and fractured ribs attested.

On entry into prison, Rosa tested positive for HIV—she had never gotten tested during her years on the streets, more out of lack of interest than fear. While Rosa knew full well that AIDS was prevalent in the drugs-for-sex communities, her indifference was not uncharacteristic of those who inhabited her world.

"I knew I could catch the virus," Rosa later told a St. Clare's pastoral care intern. "For me," she calmly continued, "AIDS wouldn't've been that big a deal. It would've been just one more bite out of the old shit sandwich. It would've been just another problem, that's all." Indeed, Rosa was echoing a sentiment typical of many Spellman patients.

Preliminary blood work in prison revealed her T-cell count was already low at 178, but her suspicions toward everyone, especially those in authority, made her refuse any further tests or follow-up medical care in prison. Because she was feeling all right and had no obvious complications of AIDS—her T-cell count under 200 automatically gave her this diagnosis—the prison doctors left her alone and never encouraged her to consider any treatment, not even the three-times-a-week Bactrim tablet that can prevent PCP. Instead, in prison, Rosa still maintained her appetite for heroin or cocaine, which trickled in through a few of the guards, who would supply the inmates with drugs in return for sex, money, or blackmail payment.

Released from prison, Rosa was quickly back in the crack houses, but she had no immune system left. Her T-cell count was zero. As often seems to happen with inmates with AIDS, Rosa started getting sick almost as soon as she was released, first with fevers and drenching night sweats, and then with an insidious, dry cough that made breathing difficult. In fact, she was barely able to walk up the short flight of stairs to the crack houses' bedrooms, where she did her "business." As a result, her business began to dwindle. The fevers, sweats, and exertional breathlessness she could largely hide from her tricks, but her uncontrollable coughing fits started scaring away clients and soon dried up her only means of getting drugs. A crack buddy of hers, a male hustler with AIDS, told her she needed some Bactrim pills for the cough. However, her old trick at the pharmacy, who had purloined antibiotics for her several years ago, was dead—from TB, according to a mutual acquaintance. Through another "friend," a Cabrini Hospital nurse, who demanded her last ten dollars in return, Rosa was able to get a fistful of Bactrim tablets, which reduced the fevers and cough, at least for a week or so.

Becoming too weak to steal and too sickly to attract even the most desperate johns, Rosa turned to panhandling on the streets, which netted her only a fraction of what she needed for her habit. Because she had lost her room at the halfway house—she had been AWOL too long in the crack houses—she had to live on the streets, not a bright prospect, given the harshness of New York's winters. Only one night did Rosa stay in one of the city's notorious homeless shelters, where she was roughed up and robbed of her small bag of clothes and cosmetics, her only remaining possessions. Ironically, for someone as vulnerable as Rosa, the streets were safer than the city's shelters. But the streets were snow-strewn and bitterly cold, and her fevers and cough were returning.

"I didn't know what to do," she recounted many weeks later to Abby A., who was trying to find housing for her when it looked as if she might actually make it out of the hospital. "I couldn't get welfare 'cause I forgot to see my parole officer. I was too messed up to do anything, and I knew they'd send me back to Greenhaven, or worse. I couldn't find my sister—she must have moved or something—and even if I did find her, she'd probably tell me to get lost, like she did before."

Thus, as Rosa's condition rapidly deteriorated and the snowbanks accumulated, her "home," by default, became the New York City subway system, where the homeless, especially the unwashed and the most deranged appearing, can almost always find a seat, even an entire row of seats. As a rule, other passengers grudgingly give wide berth to a pungent pile of soiled clothes covering a sleeping schizophrenic or crackhead stretched out on the seats of the city's subway cars, and Rosa became one of these people that straphangers studiously ignore. Dressed in dirty jeans, red sneakers, and a long black coat she had found in a trash bin, Rosa wrapped herself in an old blanket and panhandled mainly the IRT number 1 and 9 lines, which service all the local stops in more desirable areas of Manhattan. Sometimes claiming to be a homeless mother of five hungry children, other times professing to be HIV-positive and in need of money to buy medicine, a glazed-eyed Rosa would imploringly, impatiently, shake her paper cup of change in passengers' faces as she staggered from car to car.

It was not exactly clear how Rosa finally landed in the Seventh Avenue subway tunnel, a few hundred feet from the Fiftieth Street stop. Rosa could later recall nothing. Perhaps she fell off a train as she was walking between cars on her endless circuit of panhandling, or perhaps she merely sought refuge in a tunnel corner that was warmed by a nearby heating vent. A motorman on a passing train saw her motionless body off to the side of the tracks and radioed for help. Per MTA protocol, both local and express tracks on the uptown subway service were shut down while the police investigated. To those more fashionably dressed than Rosa, it was yet another frustrating disruption of subway service by yet another homeless addict. Ironically, after repeated rebuffs to her begging, Rosa M. could no longer be ignored.

It could have been a scene out of Victor Hugo's *Les Misérables,* only set not in Paris's infamous sewers of the nineteenth century, but in midtown Manhattan in the twilight of the twentieth. The two transit cops who approached Rosa in the tunnel thought she was dead, since she did not stir when they first called out to her. Sprawled on her side and partially covered by a wet blanket, she was shirtless and had on only one tennis shoe. Her face was resting in a small pool of mud and vomit, and she reeked of feces, which were oozing through the seat of her faded blue jeans. A small cut, with minimal bleeding, was on the side of her head, near the temple. Lying nearby was her paper cup, still containing her booty from the evening—$1.15 in small change, plus two subway tokens. As the cops applied their flashlights to this tableau of squalor, it looked to them like yet another medical examiner case, another homeless victim to add to the piles of unclaimed bodies already stacked like cords of wood in the city morgue.

The cops were wrong. Awakened by the glare of flashlights and the intrusion of voices, Rosa pulled the blanket over her head, loudly cursing them to leave her alone. When they persisted in demanding she move on—the police loathed scenes like this, but the subways were backing up nearly to Chambers Street by then—Rosa haltingly struggled to her feet and immediately fell on her face. She promptly had a generalized seizure, convulsively jerking in demonlike spasms that gradually subsided after a minute, as the cops passively observed the spectacle. By the time Rosa slumped back into her puddle of muck, the EMS squad, which had also been summoned with the police, arrived on the scene, and methodically, obligatorily, scooped Rosa onto their portable, collapsible stretcher, hauling her off to the waiting EMS truck and thereby, mercifully, unclogging the IRT uptown local and express lines.

Red lights flashing and sirens wailing, more as prescribed routine than heralding any emergency anyone would really care to know about, the city ambulance made its desultory trek through the snowy two blocks to St. Clare's, the nearest hospital, ferrying an unconscious Rosa to what would become her final home. From the evening shadows near the nondescript ER entrance welcoming Rosa's ambulance, it was possible to peer down Fifty-second Street and make out the glow of pale lights bathing Rockefeller Center, Radio City Music Hall, as well as scores of restaurants, where, as Rosa was being disgorged from her ambulance, limousines and taxis were disgorging their passengers for an evening's repast.

Unceremoniously—after all, they had done this many times before—the ambulance crew quickly wheeled Rosa into the small treatment area of the ER, where the nurses dutifully transferred her still-limp body to a gurney. They removed her soiled jeans, started an IV in her arm, and checked her vital signs: she was febrile at 102 degrees and had slightly labored, stertorous respirations. The ER staff, too, had initiated such care over the years for many patients like Rosa. In fact, in the early 1980s, when St. Clare's actually went into bankruptcy—even then a victim of health-care competition—the steady ER stream of "unwanted" patients like Rosa kept the hospital from closing, buying time for the archdiocese to resurrect it. Rosa's lack of ID likewise did not faze the ER staff. A cursory inspection of her thin, malnourished body confirmed their suspicions that she was just another homeless Jane Doe. The fresh needle tracks on her arms and legs, her matted and lice-infested hair, the scars and bruises on her arms, the dirt caked under her fingernails, the empty crack vial clutched in her left hand, her rotting front teeth, her terrible body hygiene—this picture signified a bleak life not unlike that of other St. Clare's patients, who would not infrequently list the Port Authority Bus Terminal or Penn Station as their home addresses.

Equally routine was the initial evaluation and treatment the ER staff methodically began as soon as Rosa was fitted into her hospital gown and was hooked up to oxygen and a heart monitor. Baseline blood work was quickly drawn, and in rapid succession, the nurses pushed IV injections of dextrose (sugar water, in case her blood sugar was low), thiamine (a vitamin, in case she had alcohol intoxication), and Narcan (an antidote to possible heroin overdose). The X-ray department was alerted to expect Rosa for a chest X ray and head CAT scan, to evaluate her fever and seizure. As these familiar chores were being carried out, the atmosphere was quiet and businesslike, perhaps even a bit jaded, for the staff had been through this routine many times before.

Then, almost predictably, or, at least to no one's surprise, Rosa suddenly awoke and bolted up in the cart, confused and frightened by the strange ER surroundings. Ripping out her IV, pulling off her oxygen tubing, and tearing off the heart-monitor leads, she tried to climb over the raised railings on her cart, all the while loudly cursing and striking out at the nurses who tried to keep her from hurting herself. Despite the barrage of profanities, Rosa was too weak, too exhausted, to sit up for more than a few seconds, let alone pull herself off the ER gurney. When the nurses tried to reassure her, to explain what was going on, she viciously cursed at them again and fell back onto the cart, angrily pulling the sheet over her head, demanding to be left alone.

Rosa presented the ER staff with a classic St. Clare's dilemma. Severely debilitated by malnutrition and totally spent from drug withdrawal, she could not walk, or even stagger out of the ER, but she still had enough life in her to keep the staff at bay. Curled up on her side in fetal style and enshrouded by her sheet, Rosa was like a hibernating she-bear, content to be left alone, but lashing out at anyone daring to disturb her. It was a standoff: she could not leave on her own power, but the staff could not let her sleep off her drug stupor in the ER for the requisite several days. Moreover, her fever and slightly heavy respirations meant she might well have a pneumonia, a possibility with dangerous implications for both her and the ER staff, given the high prevalence of TB in people like Rosa. Several subsequent attempts by the ER staff to get any information from her proved fruitless, except she did impatiently blurt out through the sheets her first name and that she had HIV, a revelation the weary ER staff pounced upon as adequate rationale to ship her quickly upstairs at 2 A.M. to Spellman Unit 3A, where, on arrival, she cursed at the nurses trying to settle her into her room. In addition, she snarled insults at the hapless on-call intern, defiantly telling him where he could stuff his stethoscope.

By sunrise that first morning on the Spellman service, Rosa had already acquired the reputation of being a difficult patient, something of a record in light of the few hours of hospitalization that had elapsed. Repeatedly refusing both care by her nurses and evaluation by the bleary-eyed medical intern, Rosa greeted dawn as an inert lump hidden under the bedcovers—inert, that is, until the ward staff would venture to disturb her for vital signs, breakfast, blood drawing, or to mop up the mess from her angrily pushing over a pitcher of water left on her bedside table. Whenever awakened, Rosa viciously lashed out at her victim with almost demonic fury.

After breakfast, which Rosa threw on the floor when a nurse's aide enraged her by offering to help feed her, the Spellman medical team arrived to interview and examine her. As her physician, I had already been warned by the nurses about Rosa, but I had to try to make sense out of her vague, confusing history. The EMS report in her chart only documented, "Young adult female found unconscious, seizing in subway tunnel," along with vital signs taken by the paramedics en route to the ER. The ER paperwork was likewise unhelpful, recording only Rosa's vital signs, the IV medications given, and her abrupt awakening: "patient uncooperative, verbally abusive . . . refuses CAT scan and exam . . . unwilling to give history." The note concluded with her ticket out of the ER: "Patient HIV+, admit to Spellman 3A." The on-call intern's note was even more cursory: "Patient refused exam, will defer to Spellman M.D." Indeed, "defer to Spellman M.D." meant the buck had been passed last night as far as it could go, from the transit cops in the subway tunnel all the way to me. Along with my PA Henry R. I quietly entered her room that first morning, hoping she had finally settled down somewhat.

As soon as I softly spoke her name and gently tapped her shoulder, Rosa exploded.

"Jesus fucking Christ! Can't anyone get any fucking sleep around here? Fuck off, asshole!" Hate permeated the room and seemed to drip from every word. Wrapped in newly stained sheets, Rosa rolled over on her side, away from us.

"Rosa, I'm your doc—"

"Go to hell! Go fuck yourself! Get out!" Rosa pulled the cover tightly over her head. *"I said get the fuck out!"* The finality of this command, shouted from under the covers, left no room for discussion, and Mr. R. and I exited.

Rosa's gratuitous outburst that morning naturally angered me, but I quickly diffused my anger by reciting to myself what I called the Spellman Attending Physician's Mantra, namely, "It's not my fault, it's not my problem, it's not my job," yet all the while still caring about, still empathizing with, her situation. In my dealings with Spellman patients, I always try to walk this fine line between altruistic self-sacrifice and angry disengagement—that is, between caring too much and caring too little.

Like countless patients before her, Rosa M. epitomized the infinite *impossibilities* of AIDS care at St. Clare's, the infinite combinations that result when *impossible* patients are afflicted with an *impossible* disease in an *impossible* hospital setting.

The first several days of Rosa's hospitalization were a real mess. Care was made particularly difficult because Rosa constantly refused to come out from under her covers. She repeatedly prevented blood work or X rays, and she drove away the nurses whenever they would come in to check vital signs or change her bed linens, preferring to stay hidden under her dirty sheets. Pleas for her to bathe, change her hospital gown, or brush her teeth were met with coarse profanities. A compelling, musky aroma greeted anyone entering her room, which had become a sty from the food hurled on the floor in fits of anger. Rosa's refusal to bathe meant her body lice could not be treated, a particularly alarming prospect for the staff, who always dreaded contracting lice or scabies from patients, which had periodically happened over the years.

During these first three or four days, Rosa's condition could only be guessed by the ward staff. The robustness of her "fuck off, asshole" and the gusto with which she heaved her food trays on the floor were the only gauges of her physical condition. Mr. R. and I dutifully, patiently, looked in on her daily, to make sure she was still alive and to keep her familiar with her ward team, but she greeted our visits with either seething silence or a vile litany of profanities. Her first interactions with the staff—other than scatological expletives—appeared on the fourth day, when she started complaining to the nurses about the lack of heat in her room and the poor TV reception. Apparently disgusted by the sticky food stains on the floor, she even told one of the nurses her room was "a real fucking dump," oblivious to her complicity in the creation of this noisome environment.

By the fifth day, Rosa and I had our first conversation—of sorts. As always, I greeted her on rounds without forced pleasantness. And, as on prior visits, she was in bed, under the covers and on her side, with her back toward Henry and me. But instead of cursing, she said nothing in reply to my hello. I assumed she was in her silent-but-seething mode. When I persisted and asked her how she was doing, her answer, although annoyed in tone, was different, and calmer, from previous ones.

"Why do you fucking assholes always ask me the same fucking questions every fucking day?" For once, there was less anger and merely weary impatience in her manner.

"I'm your doctor, Rosa, and I have to ask you how you're doing," I said with my best Ohio reserve. My voice was as neutral as I could make it.

"You assholes don't have to do fucking jackshit for me. I never asked for all this hospital shit. Just leave me fucking alone." Gone was the raw rage and hate. Sounding more frustrated than angry, Rosa was beginning to state her opinions rather than issuing commands.

Willing to push her a little further this time—after all, she had not yet told Henry and me to "fuck off"—I ventured, "Okay, we'll leave, but just tell me if there's anything we can do to make you more comfortable today." There was a brief silence, followed by Rosa's unorthodox reply.

"Nothing, unless you assholes can get me four vials of crack." The sarcasm in her voice was refreshingly new. I even imagined she might be forming a contemptuous smile under the sheets, immensely pleased with herself at her witty retort to my serious inquiry. Although her anger had not dissipated, this crude repartee denoted a break in her demeanor. Henry and I smiled at each other.

"I'm afraid crack isn't in the hospital formulary of approved drugs, Rosa," I replied jokingly.

"What a fucking shame. Then I suppose you assholes had better show yourselves to the door." Rosa pulled the covers tightly over her head, a signal that the day's audience was terminated.

From this "conversation" grew the first signs, however halting and intermittent, of Rosa's cooperation with the staff.

Rather than automatically dumping her meals on the floor, Rosa now started to pick at her food. If she liked it, she would eat it, and if not, only *then* would the tray be overturned on the floor. She began to let the nurses take her vital signs, *provided* she was not asleep, busy watching TV, or eating. Although she still refused to talk with Abby A., she did tell Sister Pascal to fetch some cigarettes. She finally consented to showering and applying Kwell lotion for her lice, but when the nurse tried to change her contaminated bedsheets afterward, she had a fit, for reasons known only to her, and chased the poor man out of the room. She also agreed to use her oxygen, but when she insisted on smoking in bed, the oxygen tank had to be wheeled out of her room. Some days she would ignore the housekeepers mopping her food-stained floor, whereas other days she would angrily order them out. Because the fever and cough noted her first night in the ER raised the possibility of TB, she had been placed in a private room and put on respiratory isolation. Staying put in her room was not a problem, at least early on, when she rarely even ventured out from under the covers, but she repeatedly ignored the nurses' requests for sputum specimens for TB testing, preferring instead to use the specimen cups as ashtrays. During those early days, Rosa kept the staff guessing as to what her response to them would be.

Rosa's feelings during this difficult period were unknown, since, at the time, she shared them with no one. Three weeks later, however, after she had recovered from drug withdrawal and had made peace with the staff and herself, she recalled for me her thoughts those first few days on 3A.

"I didn't want help from nobody. I wanted only two things—to be left alone and to get some crack. I wanted to get better, I didn't want to die. All I wanted was another hit, that next high. I can't remember much more than I wanted crack more than anything else, but"—here Rosa revealed what it was that had kept her from walking out once she had regained the strength to do so—"I was scared to leave 'cause I knew they'd get me for parole violation, and I couldn't take another term in Greenhaven." Rosa's drug binge after release from prison had caused her to miss her weekly appointments with her parole officer, and even if she had shown up, she knew that the urine drug tests would have been enough to send her back to prison.

"I had it all planned out," she later went on to tell me. "All I needed was to lie low till I got strong enough to leave New York, but then my cough came back. And when my breathing got rougher, I knew I couldn't leave. I knew I needed more Bactrim." Rosa had discovered a terrible truth about AIDS: a strong will is ultimately no match for the HIV virus.

This resurgence of Rosa's breathlessness and cough, which became a concern on the seventh hospital day, had previously gone unnoticed by Henry and me, mainly because of her nearly continuous submergence under her blankets. It was the unit's head nurse, Shirley T., who first noticed the dry cough and heavy breathing as she helped Rosa walk from bathroom to bed. Receiving this report—Ms. T.'s observations were always accurate—I questioned Rosa about these symptoms on rounds that seventh day, risking yet another tirade by actually *insisting* she consent to an examination. But, testimony to how very sick she was, Rosa relented—after the weakest protestations—and let me briefly examine her for the first time, a full week after her admission. The annoyed, impatient expression on her face during my exam belied her fears that maybe she might not live long enough for that next cocaine high.

"I was scared, scared shitless. I thought I could beat the virus," she recounted two weeks later to Shirley T. "But *I wasn't stupid.* I knew I had to let you guys help me, even if it pissed me off."

On this first exam, I found fairly recent needle tracks, thrush, raspy breath sounds, a slightly enlarged liver, fairly severe muscle wasting, and—most significant—a definite, albeit subtle, increase in her respiratory rate, as well as a spasm of dry coughs whenever she would breathe in deeply. Rosa was pulling for air, a harbinger of serious problems to come, especially if she continued to refuse further evaluation and treatment.

My cursory exam completed, I stepped back and pensively said, "You know, Rosa, I'm worried about your breathing. It seems labored, and that cough wor—"

"My breathing's fine, there's nothing wrong with it," Rosa snapped, her annoyed tone unable to conceal the first trace of worry I had heard in her voice since her admission.

"Just the same, because of your fevers, I think you need to get a chest X ray to be sure you don't have a pneumonia." The calm, even tone in my voice hid my frustration at her prior refusals to go down for an X ray. "It'll take only twenty minutes to go for the X ray."

"I don't need no fucking chest X ray. I'll be fine. It's the goddamned dry air in this room that's fucking up my lungs and making me cough. I'll be fine." Again, there was the slightest tentativeness in her voice. She seemed to be trying to reassure herself, not me, that she was okay.

"I hope you're right when you say you're okay, but think about maybe getting an X ray." I knew that my best chance of getting a chest X ray was if Rosa initiated it.

Later that same afternoon, acting as if she were doing everyone a big favor, Rosa imperiously informed Ms. T. that she would go down for a chest X ray, which showed an extensive pneumonia in both lungs. When Henry and I tried to discuss these results with Rosa later that same afternoon, she refused us audience.

"I already talked with you today, I got your fucking X ray, now get the fuck out." There was less vitriol and more impatient fatigue in her command. Rosa was almost *pleading* to be left alone, and I decided she had made as much progress as we could expect in one day. Besides, I reminded Henry and myself, AIDS is usually a disease in slow motion: her pneumonia had been festering for a while and would probably not kill her by morning rounds the next day.

By the next morning, one week and a day after her inauspicious arrival on 3A, Rosa was visibly sicker. Her temperature was 102 degrees, her hacking cough was more persistent, and her breathing seemed heavier. Overnight, in the early-morning hours, she had requested oxygen for her breathing, and after confiscating her matches and cigarettes, the nurses wheeled the monster-sized oxygen tank back into the room, hooking her up to it again. On rounds later that morning, Rosa looked weaker. Although she still sported her previous put-out attitude as I examined her, the trace of worry I'd noted in her voice the day before was now also evident in her eyes.

"It's the fucking dry air in this fucking room, I tell ya, Doc." It was the first time Rosa had addressed me as "Doc" rather than "fucking asshole." This minor triumph was not lost on me, despite the urgent business at hand.

"You gotta get me a humidifier. It's not—" A paroxysm of stuttering, hacky coughs cut short her protestations. Once she settled down and caught her breath a half-minute later, she worried out loud, "It must be the dry air, or something I'm allergic to in the fucking food. This place is such a fucking dump that—"

"Rosa," I interrupted, "I think you may have PCP, the AIDS pneumonia." As with many Spellman patients, Rosa's years in prison had educated her in the jargon of HIV medicine. She knew that PCP was not a good thing to have. For once, she fell silent and listened to me, without interrupting.

"We need to start an IV on you," I continued quickly, fearing her attentiveness might not last long. "We need to start you on Bactrim intravenously. It's the only good treatment for PCP, especially when it's as bad as yours. We *could* give you pills, but they're not as good as the IV treatment. You really need the IV."

I realized there was little chance, if any, Rosa would acquiesce to IV therapy. She was still adamantly refusing any blood work, so undoubtedly she would not stand for the added discomfort of an intravenous, especially since it would have to be changed every three or four days, if not more often. Indeed, up until then, Rosa had even been refusing to take any oral medication, claiming the staff was trying to "poison" her. Thus, framing the debate on my terms, I was presenting her with two options, IV or oral medication, in the hope she would relent enough to permit the oral therapy.

"I really recommend the IV over the pills, Rosa," I repeated, with enough seriousness to hold her attention.

"I don't want no fucking IV. I'm no fucking guinea pig. You people don't know what the fuck you're doing anyway." More weary frustration than anger was in her voice, and another brief coughing jag cut short her diatribe.

"If you won't take the IV, then I guess we have to respect your wishes and try the pills instead," I countered, trying to make her feel she had struck a hard bargain with me. "And to help the pills along, we'll order a humidifier for your room, okay?" In the brief silence that followed, Rosa seemed to be weighing the pros and cons of cooperating with this persistent doctor and PA.

"Okay," she replied reluctantly, "but make sure it works. This place is such a fucking dump, half the stuff here is crap."

Later that afternoon, courtesy of Sister Pascal and her pastoral care "slush fund," Rosa got a humidifier, a room amenity usually unavailable at St. Clare's. Miraculously, that same afternoon Rosa M. took her first medication, two Bactrim DS tablets, which, except for occasional lapses, she continued to take regularly, every six hours, over subsequent days and weeks.

Despite the substantial strides made by the beginning of the second week, Rosa M.'s conflicts with the Spellman staff persisted, at a different level. In fact, some of the staff began to yearn for the simpler, less complicated Rosa M. of earlier days, when nurses simply had to listen to vulgarities issuing from beneath a mound of dirty covers in her bed. As her appetite improved, Rosa became increasingly critical of the hospital's food, and whenever the meal was not to her liking, she would stand in her doorway and, defying the rules of respiratory isolation, loudly shout her displeasure down the hallway to the distant nursing station and to the other patients. When the nurses would thereupon chide her about violating respiratory isolation and ask her to go back into her room, she would growl out an expletive. "You're all fucking trying to starve me!" was her usual refrain, and she would then forcefully slam the door behind her. Although she did take the majority of her pills—the Bactrim tablets, plus medications for thrush, fever, constipation, and anxiety—the nurses would still find occasional pills scattered under her bed, in her bedsheets, or in the wastebasket. Her new willingness to allow Henry R. and me to examine her remained exclusively on *her* terms: some days she was "too tired" to come out from under the covers, and other times she just did not "feel like it." Even more frustrating were Rosa's tirades about problems we had no control over: the lack of hot water in her bathroom sink, the sluggish flushing of her toilet, the loose toilet seat, the erratic radiator heat, the mice scurrying under her bed at night, the inoperable electric bed, the room's inside doorknob that invariably fell off—all of these routine gaps in St. Clare's accommodations were *my* fault in Rosa's mind. She could not adjust to the fact that this was not a Holiday Inn.

As if these perennial problems with St. Clare's crumbling infrastructure were not enough to enrage Rosa—making her medical care even more problematic—there were also the time-honored deficiencies with various ancillary departments involved in her care. Several days after convincing Rosa to take Bactrim, Henry was finally able to cajole her into allowing blood work for the first time, but no sooner was it drawn than the lab lost the specimen. It was another week before Rosa would permit another blood drawing. Rosa's seizure in the subway still necessitated a head CAT scan, and about the same time she first acquiesced to blood work, she reluctantly agreed to go down to X ray for the scan. However, her one-hour wait in the drafty corridors of the radiology department did not sit well with her, nor did the radiology staff's apparently dismissive attitude toward her plight, so she stormed back to the floor before the scan could be done, seething and vowing never to go for an X ray again. Several delays in the pharmacy's delivery of Ativan, her "nerve pills," to Unit 3A—an endemic problem for Spellman floors—made her vengefully unwilling to take her other, more important medicines until her nerve pills arrived. But the hospital lapse that was most incendiary, the slight that provoked the most rages and the heavings of meal trays, was the regularity with which her daily menu selections were ignored by the hospital kitchen. A deafening "You're all fucking trying to starve me" was a regular refrain heard from her end of the hallway at mealtimes, punctuated by the violent slamming of her door and the crash of a full tray of food thrown against the inside of the door.

Nonetheless, two weeks after her arrival at St. Clare's, Rosa M. was starting to respond to the Bactrim. Her fevers were down, her cough had abated, and her strength and breathing were on the mend. Although her chest X ray a week earlier had revealed a pneumonia, it was still unclear what kind of pneumonia it was—PCP, regular bacterial pneumonia, TB, or any combination thereof was possible. Indeed, Rosa's case was an example of how AIDS had profoundly humbled medical science. Evaporating any previous veneer of diagnostic certainty, HIV infection often obscures previously well-defined diseases such as TB or bacterial pneumonia, resulting in subtle, atypical presentations. Further complicating diagnosis is the alarming frequency with which several different opportunistic infections can coexist simultaneously in a patient, resulting in vague, unusual symptoms. Thus, diagnosis of a pneumonia such as Rosa's could well prove elusive. No sooner would the doctor feel the diagnosis was cinched than everyone would be blindsided by another, unsuspected microbe causing identical symptoms. Often, multiple and repeated tests would be necessary before finding the cause, or causes, of a pneumonia—or of a diarrheal illness or a rash or a headache or a fever.

When, as in Rosa's case, the patient refuses diagnostic tests, the task of medical evaluation can be nearly impossible. The result is that AIDS care, especially at an AIDS facility in an inner-city setting, has restored emphasis to the "art" of medicine. Clinical intuition and judgment, those intangible skills making a good doctor more than just a "body mechanic," are often more useful than a whole array of tests and X rays in pinning down the chimera of AIDS. Rosa, of course, refused bronchoscopy to evaluate her pneumonia, and only after much coaxing did she finally provide three sputum specimens, which were negative for TB, thus releasing her from respiratory isolation. Rosa's Bactrim pills would treat both PCP and most bacterial pneumonias, an inelegant but practical solution to her recalcitrance toward more definitive evaluation.

As important as clinical intuition in AIDS care—and in Rosa's case, even more important—is emotional support to those afflicted by this stigmatizing disease. For Rosa, as with all AIDS patients, treatment required that one look at her as more than just a body with symptoms and physical findings. Key to her medical care was the awareness that there had to have been a past Rosa M., a history that preceded her descent into the Seventh Avenue subway tunnel. Initially, of course, none of the Spellman staff knew much about Rosa's past; only later did she begin to talk with the staff. Yet despite the myriad frustrations—despite the countless "fuck off" salutations and ear-wrenching slammed doors—I perceived there was once a little girl named Rosa M., somebody's daughter or granddaughter. Daily, on the subway or on the streets, I would see beatific little Hispanic girls, clinging to a relative, appearing much as I liked to think Rosa once had. Though it may sound maudlin, such ability to project, to imagine this snarling patient as a little girl, became imperative to her treatment.

It was difficult to pinpoint exactly when Rosa's soul began to resurface from her private underworld, but rumblings were detected sometime between the second and third week of her hospitalization. Perhaps it was the first time she found her meal not to her liking and did not push it onto the floor, instead simply opting to complain about it to her nurse. Or maybe it was the first time she held the door open for a housekeeper struggling to take his mopping bucket out of her room. Significant were the first times she said "thank you" to her nurse for bringing her clean towels or to Henry and me as we finished visiting her one day. Helping her nurse make her bed, appearing interested when Sister Pascal dropped off some magazines, agreeing to talk with Abby A. about housing needs—all of these glimmers of humanity coincided with fewer tirades, fewer glaring stares, when something went wrong. For Henry and me, daily rounds on Rosa became more civil: she started asking questions about her condition, sometimes as challenges, but more often as concerned inquiries. Blood drawing on her was no longer a hassle, and in a major breakthrough, she willingly endured another wait in the inhospitable corridors of the radiology department to get her head CAT scan, which was fortunately normal.

What was going through Rosa's mind during this transition period was unclear, since she still said little about her feelings. Although more alert and cooperative, her demeanor remained quietly aloof and suspicious, as if she expected the staff to put her down, to treat her as dismissively as she had been on the street. Ever observant, she intently eyed everyone who came into her room, silently watching every move of the nurse making her bed or handing out her medicines, vigilantly following my every motion and word on daily rounds.

"At first," Rosa later confessed to me, "I didn't know what to make out of this place, but after a while, I saw this place was different . . . you guys weren't like all the others."

As Rosa was marking time for her escape from the hospital and New York, she started thinking about her almost indescribable descent into the netherworld of New York. St. Clare's was a radically different venue for her remembrances, and this quirky little hospital reminded her of her last real home, her grandmother's house, where she had received unconditional love for the last time.

"After a while, I'd lie in bed at night, thinking about what a piece of shit I was," she later told the Jesuit seminarian in the pastoral care program. "But I never stopped believing in God. I *wanted* there to be a God so He could punish me for all the terrible things I'd done. I guess I wanted everyone here to treat me like shit, but whenever I'd tell someone to fuck off, they just kept on being nice to me. I didn't get it: the shittier I was, the nicer they were to me."

Once her meal trays stopped ending up on the floor, Rosa started putting weight on her 104-pound frame. Her complexion became less sallow, her face less hollow. One morning, close to her third-week anniversary, Rosa shocked everyone on the unit. Her long, straggly hair, which had been matted ever since her self-imposed subway incarceration, was not only cleaned but also carefully braided back into a bun, which was highlighted by a bright red barrette. Fresh makeup, which Sister Pascal had brought her the day before, accented her jet-black hair, with striking effect: Rosa M. appeared luminous. On rounds that day, Henry and I said nothing to her about this transformation, so as not to appear patronizing, but Rosa nonetheless detected our surprise. Although likewise not acknowledging her "new look," she appeared intensely pleased with herself, as if to say to the staff, "You thought I didn't have it in me, didn't you?"

By the middle of the third week, Rosa started venturing out into the hallway, cautiously exploring the heretofore unknown landscape of Unit 3A. To their caregivers, previously bed-bound patients always appear different once they are back on their feet, and it was especially strange seeing Rosa's lanky body tentatively navigating the 3A corridors. On these initial forays out of her room, Rosa seemed quietly curious, and on one of these walks past the nursing station, she smiled at me for the first time.

"Hi, Rosa," I greeted her, "it's good to see you on your feet. I'm glad to see you're taking it slow and easy to start."

"Oh, I'm just checking things out. I want to see where you guys hang out," she replied nonchalantly, even playfully. "I used to walk a lot with my grandma." Here her voice was suddenly hesitant, as if she were recalling a distant, long-repressed memory. "Yeah, I used to walk a lot, when I was a little girl." Her wistful tone confirmed what I had long suspected: there had indeed been a child named Rosa M.

Although medical science deserved credit for Rosa's recovery from pneumonia, it was Abby A. and Sister Pascal who made Rosa feel comfortable enough to talk about herself with others. Abby was actually the first person to whom Rosa opened up.

A progressive, noncloistered Catholic sister in her forties, Abby had not been scandalized in the least by Rosa's initial rebuffs gratuitously laced with profanities. Although of gentle temperament, Abby had always had a strong sense of self, of her emotional boundaries, a trait essential to being a good social worker on a Spellman unit. By the third week, Abby and Rosa finally started talking, and after two stiffly formal visits, when Rosa seemed to be sizing up her social worker, the subject somehow changed from housing needs to Rosa's desire to reestablish contact with her sister, Maria.

"I haven't seen Maria for years. She's all the family I got, except for relatives in Puerto Rico I don't know. I tried to call her when I got paroled, but she'd moved—where to, I don't even know. She probably won't want to see me. I treated her like shit right before I left. I don't know why I want to see her, but I'd like to." Initially indifferent in tone, Rosa's voice trailed off, and she appeared pensive.

"Your parents aren't alive, then?" inquired Abby.

"Nah, they died in a fire when I was thirteen, but I always lived with my grandma, but then she died right after that, and I went to live with my sister. We never got along." Rosa's voice seemed to quaver, and she turned and nervously started to leaf through a magazine, pretending to read it.

"Just one more question, Rosa, and then I'll come back tomorrow." Abby sensed Rosa's uneasiness and did not want to annoy her. "If I find your sister, is there anything you want me to tell her? I mean, if she asks, should I say anything about your medical condition, or should I just tell her you're here and let you tell her whatever you want?" Rosa put down the magazine and, in her first expression of remorse since admission, looked very, very sad.

"If you find Maria, tell her I got AIDS and it's God's punishment for what I did to Mama Rosa." Rosa's reddening eyes were brimming with tears.

"I don't understand. What punishment?"

"When Mama Rosa—my grandma—passed, I felt guilty, I felt I did it, I thought Mama's death was my punishment from God." Rosa wiped her tears on her bedsheet. "And now that I've got AIDS, I *know* I'm being punished."

"Punished for *what?*"

"I hated my mother and father, *hated* them. I know it's awful to say, but it's true. They were always drinking and yelling at Mama. 'You're too old, we can't take care of you anymore, why don't you move,' they told her. Once Daddy got real drunk and yelled at Mama and said, 'Why don't you just die, old lady?' I *hated* him for that. I wanted *him* to die.

"When Mama moved out, I went with her. My parents didn't give a fuck what I did." The tears were pouring forth, but Rosa seemed determined, almost compelled, to speak her feelings nonetheless. It was as if she were powerless in holding back any longer.

"When Mom and Dad died, I didn't feel sad—I was *happy* they were dead. I remember even thinking at their funeral, 'I'm glad they're dead, now Mama and I can live happily ever after.' Then Mama died right after that, and I didn't know what to do. I knew God was punishing me for my bad thoughts—my bad thoughts killed my Mama. Now that I've got AIDS, I'm glad I'm going to die . . . I deserve to die."

Having grieved with countless troubled patients like Rosa, Abby knew it was best to say nothing right then about Rosa's self-blame, to resist the urge to say "the right thing" in a misguided attempt to assuage her hurt. Such "happy talk" would only trivialize Rosa's pain and convince her that no one really understood. Abby sensed that Rosa's healing could come soon, that her confessions that day were a beginning as important as when she agreed to get a chest X ray and swallow her first Bactrim pills several weeks earlier. Abby knew it was best to refocus Rosa's attention on more neutral subjects.

"Where did you and your grandma Rosa live?"

Rosa's spirits picked up at Abby's question. Eager to change the subject of her childhood somewhat, but not entirely, Rosa launched into an animated description of her days with Mama Rosa—the trips to the market, the regular visits to church, and the long private walks they made along the Grand Concourse, her window onto a tantalizing urban world.

"I know it doesn't seem like much now—you know how little kids think some place is so great, and later on, when they get older, it doesn't seem like such hot shit—but the Concourse was 'the big city' for me. With my Mama, I felt like *somebody* walking down the street in a pretty dress she'd made for me."

"You know," Abby said, "my aunt used to live off the Concourse, on One Hundred and Seventy-fifth Street, and we'd go up when I was a little girl and visit. I know exactly what you mean—it's always busy, always active."

For the next five minutes, Abby and Rosa chatted about seemingly insignificant recollections from their childhoods. Thereafter, Abby would look in daily on her friend, not only to help arrange housing, but also to mediate a new start with Rosa's parole officer and, most important, to talk about whatever Rosa wanted.

A few days later, Rosa had another important conversation, this time with

Henry R., her physician's assistant. Other than morning rounds with me, Henry's visits to Rosa's room were usually for business. Her veins had become so bad that the regular phlebotomist could no longer get blood from her, and Henry was somehow able to hit the vein on first try, a skill that immediately endeared him to Rosa, who, despite her prior drug use, hated needles, grimacing in pain like a ten-year-old whenever she got stuck. After the stick was finished, Henry would usually linger for a moment, to talk briefly with his blossoming patient, usually about innocuous things, such as the oil painting she had been working on in the afternoon crafts sessions, or her application for admission to Gift of God, an AIDS boarding home. However, a couple of days after her talk with Abby, Rosa's conversation with Henry grew more profound, once the postphlebotomy pleasantries concluded.

"Henry, do you believe AIDS is ever a punishment from God?" Rosa's inquiry sounded sincere, not contentious. Never talking much about himself to anyone, Henry was a very private young man who had strong spiritual beliefs founded not only in religious faith, but also in an incredible knowledge of philosophy and history. Indeed, Henry R. was the most intellectual PA I had ever worked with.

"No, Rosa, I don't believe AIDS is ever a punishment from God," he calmly replied. He betrayed no hesitation. "We punish ourselves by failing to learn and grow from our mistakes. Our mistakes in life occur when we don't follow the example set for us in Christ." Gathering up his blood-drawing paraphernalia and the tubes full of Rosa's blood, he concluded assuredly, "God loves us no matter what we do, and He wants us to love Him."

"But some people say AIDS is a punishment from God," Rosa tentatively countered, more as a challenge to herself than to Henry, who was on his way out the door, to his next ward chore.

"AIDS is our opportunity to love one another, as God loves us. Think about it, Rosa. I'll see you tomorrow." Henry's innate sweetness, his total lack of hard sell about his beliefs, reassured Rosa that the conviction in his words was genuine. Henry's centered, but not self-centered, demeanor was, Rosa thought to herself, how *she* wanted to be, despite her many mistakes of the past. Gradually, Rosa resolved to emulate her caregivers on 3A.

Rosa began subsequently to question her feelings that AIDS was a punishment from God. Sister Pascal's pastoral care visits gave her the chance to think aloud, to forge for herself the connection between her Mama Rosa's love and her present tumult, a tumult that ranged from guilt about her parents' deaths to her fears about God's infernal wrath. Always the quiet listener, Pascal implicitly understood the redemptive power of resurrecting Mama Rosa's love.

"I remember my first Communion," Rosa related to Pascal, "and how happy Mama was. Mom was working, and Daddy overslept—he never went to church anyway. So it was just Mama, Maria, and me. Mama gave me a new Bible, and in it she wrote, 'I'll wait for you in heaven, we'll walk with Jesus,' or something like that. When I was on the streets and into drugs, I tried not to think about Mama, but when I did, I felt there was no way I'd ever see her again. I'd done so many bad things, there was no way I'd ever see her in heaven." Rosa spoke steadily, without trembling, but with a recognizable trace of remorse.

"But now I want to see my Mama so bad I can almost *feel* it, Sister." There was a pleading, almost reverent tone in Rosa's words. "I feel her love all around me here—with you, Dr. Baxter, Abby, Shirley T., Henry . . . I know it may sound crazy, but I can feel my Mama real close to me."

"It's not crazy at all," Pascal said, smiling at Rosa. "Your Mama's love came from a higher source, and take it from me, that sort of love never dies. It didn't die with your Mama and it didn't die even when you turned your back on it."

With a trace of the Bronx in her cadence, Pascal continued, "Now that you've once again embraced that love from your Mama, you're sharing it with everyone here. Your Mama's love is still alive."

Out of respiratory isolation, Rosa was moved to a semiprivate room on the unit, where her roommate was Yolanda J., a twenty-year-old black woman who, emaciated and only intermittently conscious, was in the final stages of metastatic cancer of the cervix. Except for the few hours each day the nurses put her eighty-two pounds of skin and bones into a bedside lounge chair, Yolanda was bedridden and unable to care for herself. At first, Rosa had eyed her roommate warily, not quite sure how to react to her occasional requests for help—which usually were for a blanket to cover her feet or for the nurses to bring her pain shots. But just as Rosa had first carefully observed her ward team, only to conclude they were her friends, so she likewise sized up her new roommate and quickly decided that Yolanda was not only okay, but also almost a beatitude that had come into her presence.

"Everyone here helped *me* when I was sick and messed up," she later related to head nurse Shirley T., who was helping Rosa make Yolanda's bed one day, "so, I felt I should do something for someone else. AIDS is our chance to help others, and I feel Yolanda is my chance to share Mama's love."

Soon Rosa was assisting Yolanda with meals, passing her bedpan, and—when they did not respond quickly enough to Yolanda's call button—reminding the nurses of her roommate's need for help. A regular sight on 3A became that of an energized Rosa wheeling a cachectic, barely conscious Yolanda to the patient lounge for a smoke or to noon mass in the chapel or to afternoon craft classes, where a stupefied Yolanda blankly looked on as Rosa "assisted" her with the craft project, about which Yolanda, of course, had nary a clue. Rosa began assisting Yolanda with her morning toilet, brushing her thinning hair and even applying makeup and nail polish—and, to complete her ensemble, Yolanda was adorned with beaded rings, necklaces, and bracelets Rosa had fashioned in crafts class.

Yolanda had no known family—or, at least, no family Abby A. had been able to locate—and like the majority of Spellman patients, she never received the endless procession of Hallmark cards, flowers, and get-well balloons that festoon the rooms of most hospitalized Americans. Determined to correct this oversight, Rosa set about beautifying Yolanda's side of the room as best she could, hanging on the walls the two oil paintings—both of floral arrangements—she had made in crafts and, from materials in the class, creating several ornate get-well cards, which decorated Yolanda's bedside stand. Whenever slightly used flowers would arrive at the nursing station—St. Clare's was a convenient charity destination for fading flowers donated from nearby theaters and restaurants—Rosa would grab a bunch, rearrange them in a vase borrowed from Sister Pascal, and proudly present them to her semiconscious roommate, who, at most, would weakly nod her thanks. Except for requests for pain shots, Yolanda never said much to anyone—she was too far gone to have a sustained conversation—but Rosa would always chat with her as if she could understand everything. Undeniably, Rosa, in what might be called transference, saw herself only a month before in the wasted, wraithlike image of Yolanda. Rosa could identify with her hapless neighbor more strongly than any other person at St. Clare's, and if Rosa could emerge repaired from Manhattan's catacombs, so, too, she believed, could Yolanda recover. Thus, Rosa would regale her roommate with childhood stories from the Bronx, and she tried her best to keep the "conversation" from being one-sided, always encouraging—even pretending to hear—Yolanda's responses, despite the blank, dull stare of someone with HIV dementia.

"I know she probably doesn't hear much of what I say," Rosa explained to Shirley T., "but she *does* know when I'm talking to her. I like to think she understands me. I want her to feel she's being treated as a normal person." Friendly banter, Rosa believed, validated Yolanda's humanity, making her feel she was more than just another person dying from AIDS. Indeed, despite Yolanda's slow deterioration, Rosa's attention, like an elixir of ephemeral duration, did seem to perk her up somewhat. She seemed less lethargic at times and definitely ate more when Rosa fed her.

Originally transferred from a nearby nursing home for a partial bowel obstruction from her cancer, Yolanda had been born in Brooklyn twenty years before. She had been on welfare ever since birth and had been in prison briefly in her late teens, when AIDS and cervical cancer incapacitated her. Later, she had been paroled to a nursing home, where records indicated she had no visitors prior to her transfer to St. Clare's. Severely demented on arrival at Unit 3A, Yolanda could not flesh out this skeletal curriculum vitae with any further details. Nonetheless, I realized that, as with Rosa, Yolanda's now forgotten memories also once existed within a pantheon of hopes and joys—her childhood paradise, her own sad account of that paradise lost. Perhaps Rosa, in her ministrations to Yolanda, also instinctively appreciated this obvious, yet crucial insight, which is prerequisite for compassion and love.

As the day of Rosa's transfer to the Gift of God boarding home approached, she worried how Yolanda would do after her departure.

"She's so weak," Rosa said to me, safely out of Yolanda's earshot. "I know you'll take good care of her, but I worry about how she'll do at night, without me there to help." Rosa was to continue her convalescence at the boarding home, where she hoped eventually to return to school, to get her GED, and to go on to be an HIV counselor or outreach worker in her old Bronx neighborhood. Everyone assumed Yolanda would die long before AIDS claimed Rosa.

The evening prior to her discharge, Rosa spiked a 102-degree fever, which, by rounds the next morning, had returned to normal. Because she had been afebrile for several weeks, I felt such a substantial and unexplained temperature elevation could not be shrugged off, and I advised Rosa I wanted to observe her for another day or so, to be sure the fever was just a quirk, as sometimes happens in AIDS. Although disappointed at postponement of her transfer to Gift of God, Rosa nonetheless was philosophic, and perhaps a little relieved she would have to put off for a few days her goodbyes to the denizens of 3A. Besides, Yolanda was looking a little worse that morning, much to Rosa's concern.

But Rosa's fevers persisted over the next two days, accompanied by enervating fatigue and drenching night sweats. Too tired to go to craft classes and staying close to her room, Rosa was sick again, and she seemed almost apologetic to Henry and me on rounds.

"Maybe I overdid it last week, or maybe it was something I shouldn't have eaten. Greasy food sometimes upsets my stomach, and I did have pork chops a couple days ago," she explained, acting as if it were her fault her temperatures were hovering between 101 and 102.5 degrees.

"I don't think so, Rosa," I replied reassuringly, trying to conceal my worry. Fevers such as Rosa's usually presaged trouble. "Hopefully it's nothing, but I think we should get some tests, to be sure."

"Whatever you think's best, Doc," said Rosa, now ever upbeat. Perceptive enough to notice any apprehension I was trying to hide, she went on to reassure me, "Don't you worry, Doc, everything's going to work out." It is not unusual for the patient in such situations to switch roles and become comforter for the worried caregiver.

Preliminary evaluation of Rosa's fever—physical examination, repeat chest X ray, blood and urine cultures—was unrevealing, but on the fourth day of fevers, their cause became apparent. Rosa had developed jaundice, her eyes and skin turning an alarming orange-yellow hue. Nausea, right upper-quadrant abdominal pain, and complete loss of appetite quickly followed, along with a worsening of her fatigue, which confined her to bed most of the time. The only time she stirred was to go to the bathroom and to help Yolanda. Later that fourth day, initially against her protests, I moved Rosa to a private room next to the nursing station, for closer observation. Henry and I solemnly promised her we would make extra certain that Yolanda's needs would still be taken care of.

"It's just she's so weak, I'm worried about her," fretted Rosa. "I know I'm sick, but, praise God, I'm well enough to help her out when she needs it, and her flowers are wilting, and I wanted to get her some more when another delivery came. Doc, will you be sure she gets some fresh flowers when they arrive?" I promised we would.

Without delay, an abdominal sonogram was ordered, and by the end of the fifth day of fevers, a diagnosis had been made: a tiny stone was blocking Rosa's common bile duct, obstructing bile flow from her liver into her small intestine. This bile buildup not only was causing her jaundice but also was threatening serious liver damage, which could ultimately lead to bacterial infection in both the liver and the bloodstream. Such bacterial infection in the blood, called sepsis, can quickly spread throughout the body, often with fatal consequences. Moreover, as body toxins normally excreted by the liver accumulated behind Rosa's obstructed bile duct, defective blood clotting could cause spontaneous bleeding, and the neurologic effects of such liver toxins could eventually lead to coma and death. Adding to Rosa's problems was that, as revealed by blood work done on her several weeks earlier, she had at some time in the past been infected with hepatitis C, a common risk of IV drug use and a viral pathogen that renders the liver more susceptible to damage, such as was now occuring from the bile-duct stone.

Henry and I easily recognized the urgency of Rosa's condition. We knew that once hepatic failure ensued, Rosa's deterioration would be precipitous, not leisurely or incremental. Ironically, after recovering from an AIDS-related illness, pneumonia, Rosa was now facing an equally serious problem that most likely was *not* HIV-related. Gallbladder stones lodged in the common bile duct can afflict anyone. Indeed, many times in the past, I have had to use internal-medicine skills to treat non-HIV problems in various AIDS patients.

Although only five days into her new illness, Rosa already looked dehydrated, her face more wrinkled, her eyes more baggy. She seemed to have aged several years over just the previous few days. Broad-spectrum antibiotics and intravenous fluids were begun, but only as temporizing measures—more definitive treatment was imperative.

As is often the case in medicine, such definitive treatment was obvious: removal of the obstructing stone, thereby alleviating the blockage and its deleterious consequences. But, at Spellman, such no-brainers are sometimes not solved the easy way, and the obstruction was not only in Rosa's bile duct, but also in the response of Spellman's consultative services to her plight.

Late that fifth day of Rosa's illness, when the common duct stone was diagnosed, I urgently called the hospital's on-call surgery team to see Rosa, hoping it would quickly concur that immediate surgical decompression of the bile duct was necessary. Although promptly reviewing her case that evening, the surgeons—as I had feared—opted for nonintervention, a common surgical response to Spellman patients. All too often, surgeons called in to evaluate a St. Clare's AIDS patient conclude the patient is not sick enough for surgery, and "watchful waiting" is advised, or the patient is deemed "too sick" to withstand surgery. Such a catch-22 situation has often become a nightmare, when watchful waiting of a not-so-sick patient merely allows time for deterioration, the patient then becoming too sick to operate upon.

During the discussions with the surgeons that evening, I felt helpless and frustrated as they floated one obfuscating "recommendation" after another. Perhaps, the surgeons suggested, Rosa needed intravenous feedings before considering surgery, or perhaps everyone should wait to see if the stone would pass spontaneously, or perhaps the antibiotics alone might do the trick, or—the best suggestion they could come up with to get them off the hook—perhaps it would be best to call in a gastroenterologist to try an ERCP first.

A relatively new and sophisticated procedure, ERCP—initials for a word even most doctors could not pronounce correctly—involved passing a flexible fiber-optic scope through the patient's mouth and into the small intestine. The gastroenterologist would then use a small wire extension on the scope to go into the bile duct and snare the stone, pulling it down the duct, into the intestine, where it would pass harmlessly in stool. There was no guarantee an ERCP would be successful, but it was worth a try, *if* it could be arranged expeditiously. That evening I contacted the gastroenterologist covering 3A, Dr. Alma K., a private-practice physician hired by Spellman to provide part-time GI consultations. Unfortunately, some of the subspecialty consultants retained by Spellman did not project the same empathy toward AIDS patients that the full-time staff did, and on some of her prior consultations, Dr. K. had been cursory and brusque. Such attitudes had to be tolerated, since, according to the Spellman administration, most subspecialists were not interested in working for St. Clare's AIDS service. Indeed, the Spellman Center had been unable to retain a full-time infectious-disease consultant for almost a year. Replying she would see Rosa the next morning, Dr. K. was all Rosa had at that point, given the surgeons' reluctance to do anything.

Before leaving that evening, I stopped by to see Rosa again, to tell her the "game plan," not sharing with her my worries about whether this plan had a chance of success. Although considerably weakened from overwhelming nausea, Rosa listened carefully to my explanations of what was wrong with her bile duct, how serious it was, why it had to be unclogged, and how we hoped to do it. Gone from her face were the doubt and suspicions of so many long weeks ago, now replaced with trust and—to my relief *and* consternation—a sense of peaceful resignation. I could not pinpoint it exactly, but Rosa seemed to know she was imminently dying, that what was to follow in the next few days would be futile. Over and over again in my medical experience, certain patients—HIV-positive and otherwise—

would have an uncanny sixth sense about their impending demise, sometimes to the point of even telling me outright that they were going to die soon—and then dying, as they had predicted. Although not speaking of death, Rosa conveyed that evening a serenity, an otherworldly repose, that startled me. After finishing my explanation, I asked, as I always did, if she had any questions.

"Yeah, how is Yolanda? I worry about her, Doc, she really has nobody." The compassion that suffused this inquiry bespoke a miracle that, though hidden from the rest of the world, transcended all of the other quotidian events that were occurring that evening in New York. "How's she doin', Doc?" Rosa asked again.

"She's comfortable. I increased her pain medicine today, so she should sleep through the night." I usually did not discuss such details with other patients, but I regarded Rosa as Yolanda's family.

"Could you do me a favor before you leave tonight, Doc?" Rosa had never before asked me for a "favor."

"Of course, Rosa, just name it."

"Before you leave, could you please stop and tell Yolanda I'm praying for her. Tell her . . . tell her I love her." There was in her voice no trembling, only intense worry, the same fear I had heard many times before in the voices of others holding vigil for their loved ones with AIDS.

"Consider it done, Rosa," I pledged, about to be overcome with emotion. "Now, try and rest for tomorrow."

After writing a few final progress notes, I stopped by Yolanda's room. She was sleeping, having been given a pain shot only an hour earlier. Somehow, Yolanda seemed weaker, more ghastly appearing than she did earlier in the day on rounds. It was as if the loss of Rosa's companionship had withdrawn a vital prop to whatever dwindling life force Yolanda had left in her. Over the prior weeks, Yolanda's deterioration had been expected, even welcomed, by the ward staff, whose only goal had been to ease her pain until its final relief was realized. Approaching the bedside, I bent close to her ear and softly told her, "Yolanda, it's Dr. Baxter. Rosa wants me to tell you she's thinking about you and that she loves you." Neither opening her eyes nor saying anything in reply, Yolanda took in a few deep breaths right after my words, then quickly returned to her shallow respirations.

On my way home, I was filled with foreboding about what would happen.

As promised, Dr. Alma K. saw Rosa early the next morning, finishing up her brief consultant's note just as I arrived on the floor. Dr. K. fully agreed with the diagnosis and the need for emergency ERCP, which, she hastened to add on her way off the floor, she had not been trained to do.

"Can you help us get someone who does do ERCPs?" I queried with concern, calling down the hallway to my exiting gastroenterology consultant.

"I'm afraid not. Maybe the surgeons can help you."

Stunned by Dr. K.'s indifference—she did not even offer to approach a gastroenterology colleague for help—I stood by speechless as she briskly disappeared through the doorway, deftly deflecting Rosa's problem from herself. I had encountered similarly dismissive attitudes before from Dr. K., and from too many other Spellman consultants, but not when the patient's condition seemed so precarious. I immediately tried to contact the two other part-time gastroenterologists covering other Spellman units, but both were out of town, on skiing vacations.

Indeed, by that sixth morning of fevers, Rosa's condition seemed more than precarious. Her jaundice was deeper in hue, her fevers remained high despite antibiotics, and her mind seemed slower—she awoke slowly at first and was no longer as spontaneous and interactive as she had once been. Hepatic coma is not a bad way to go, I bitterly thought to myself as I quickly examined Rosa on rounds. I took solace in the fact that a week ago she had requested DNR status, obviating the gruesome ritual of CPR and life-support systems. As I was gazing at Rosa, who seemed to have aged decades even since my visit last evening, I felt overwhelmed by a nauseating wave of anger and frustration about the total impossibility of the situation. Recalling the prescient feelings Rosa had voiced the previous evening, I now knew Rosa was going to die soon, and I felt helpless to do anything about it.

"So, what do you think, Doc?" Rosa's speech was already slurred from the liver toxins slowly dulling her mind.

"I'm not sure, Rosa." I spoke evenly and slowly, trying hard to hide my worry. "I'm going to ask the surgeons to see you again. I hope they can help. Dr. K., the doctor that just looked in on you, says she can't do anything." I held Rosa's hand, gently stroking it, trying to find words of reassurance when there really were none.

Although not as alert as usual, Rosa immediately picked up on the apprehension in my voice. "Don't worry, Doc, it'll turn out okay . . . you're doing your best." I had sadly become familiar with her new refrain.

"Thanks, Rosa, you're very kind," I replied, not knowing what else to say.

"Oh, yeah, Doc . . . how's . . . how's . . ." Rosa was trying hard to remember.

"Yolanda?"

"Yeah. How's Yolanda?"

"She's just fine," I confidently answered, not regarding "fine" as a lie, since Rosa's friend was completely comatose that morning, feeling nothing. "She's not having any pain today."

"Thanks, Doc. You've been great. I'm sorry if I've been a pain in the ass—"

"Rosa," I interrupted firmly, "you've *never ever* been a pain in the ass. We all love you . . . *very much*. Now try and rest." I quickly left the room, the dry knot in my throat threatening to explode into tears if I stayed any longer, if I heard any more thanks for my "great" care.

Reconsulting the surgeons later that morning, I petitioned them to reconsider emergency surgery, emphasizing both the extremity of Rosa's condition and how surgery was her only realistic hope, given the lack of ERCP at Spellman. But, after revisiting her early that afternoon, the surgeons pronounced Rosa "too sick" to operate upon, again insisting that emergency ERCP was the only option. In rapid succession, I contacted both the assistant medical director and the medical director for help and, in equally rapid succession, was informed nothing more was available to them either. "Do whatever you can" was their collective advice. Until then, neither of my bosses had been aware that Dr. K. did not do ERCPs.

By late that afternoon, the afternoon of the sixth day of Rosa's new illness, Rosa's plight was weighing down on her ward team, and a pall of largely unspoken sadness, anger, and frustration had settled over Unit 3A. Having faced similar impasses in patient care before, the staff were already trying to comfort one another, even though Rosa was not yet dead.

Henry R. had visited Rosa several times that day, ostensibly to make sure her intractable nausea was responding to the Compazine injections he had ordered earlier in the morning. Sleeping intermittently during the afternoon, Rosa appeared comfortable to Henry, who did not awaken her, primarily because he did not want her to see how distraught he was. Rosa had always been very perceptive of people's moods. Henry felt the same frustration I did, but our despair—and anger, even rage—was not discussed that day, neither of us wanting to unsettle the other.

Like ever-present attendants to the center-stage drama, the 3A nurses fully comprehended Rosa's predicament, as well as her doctor's and PA's frustrations, and gave the three of us special attention that day. Frequently repositioning Rosa in bed and changing any soiled bed linens, the nurses promptly reported to Henry or me any evidence of discomfort, nausea, or vomiting. When discussing Rosa's condition with us, the nurses, by kind words and sympathetic glances, showed that they, too, appreciated the dilemma everyone was facing.

Abby A. likewise stopped by the unit several times that day to get updates on my attempts to get surgery or ERCP. She, too, had been through Spellman tragedies like this one before and knew both Rosa and her caregivers needed a supportive word. Moreover, Abby had some good news, of sorts. After several weeks of fruitless searching, she had finally located Rosa's sister the previous evening. Maria, Abby reported, had moved back to Puerto Rico several years ago, having changed her last name when she divorced and remarried there. Over the past week or so, Rosa had still occasionally spoken of Maria, but her recent attention to Yolanda's needs had largely displaced talk about her older sister.

"I told Maria last night Rosa was very sick," Abby related to Henry and me, "and that she'd better come right away if she wanted to see Rosa alive. I didn't tell her she had AIDS, but I felt I had to impress on her that time's running out. She seemed grateful for the news and took down our address, but she really didn't say much. So, we'll see . . . you never know." Abby went on to explain that, because she did not know how Maria would respond to the alarms about her younger sister, she did not want to tell Rosa yet, at least not until she heard from Maria again. "I just don't think we should get her hopes up, sick as she is," Abby concluded, a decision Henry and I fully concurred with.

By late that afternoon, the informal death watch was continuing. In her final throes of life, Rosa was by then barely arousable, sometimes softly moaning, and when awakened, she would only whisper a few words before falling back to sleep. Her jaundice seemed even darker, her face puffier, her respirations shallower. Her breath had the distinctively sweet odor of liver failure. Her arms and legs involuntarily twitched every so often—further evidence of advancing liver disease—and her orange-colored skin began to break out with the numerous small hemorrhages and bruises indicative of defective blood clotting. Throughout the afternoon, she had several small involuntary bowel movements, and a bladder catheter was inserted because of urinary incontinence. The rapidity of her deterioration that sixth day of illness, I surmised, reflected the lethal combination of preexistent hepatitis C, acute liver damage from the stone, and septic shock in a woman who had little, if any, immune defenses left. Indeed, my anger at not being able to get surgery or ERCP was tempered somewhat by the speed of her liver failure and sepsis: even if surgery or ERCP had been attempted, it was possible Rosa might not have survived this catastrophe. However, such speculations were specious rationalizations, since neither procedure had been done.

Uncertain whether her incoherent moans denoted pain or merely reflexive sounds from her subconscious, perhaps from dreams, I debated late that afternoon whether to begin Rosa on a morphine drip. However, my ruminations quickly ended with the unexpected arrival of Rosa's sister, who, after Abby A.'s call the prior evening, had apparently taken the next available flight from Puerto Rico. Escorted onto the unit by an evidently relieved Sister Pascal, Maria appeared somewhat apprehensive as Pascal introduced her to me, and as I solemnly apprised her of the gravity of Rosa's condition. Listening carefully to my cautionary report, Maria seemed to comprehend fully that her sister was dying. An attractive, prosperous-appearing woman in her forties—I later learned she ran a large travel agency in San Juan—Maria admitted that she and Rosa had not spoken for many years.

"I didn't even know she was in jail, until Sister told me a few minutes ago. A couple of years back, she called me late at night, asking for money. I suppose I wasn't very happy with her back then . . ." Maria's voice trailed off with unspoken regret.

"Anyway," Maria quickly resumed, with greater composure, "that was then and this is now. She's still my sister, and I came as soon as I found out. I want to help any way I can."

After warning Maria of how terrible Rosa looked, I walked Maria and Pascal to her room and left the three of them alone. Ten minutes later, Pascal emerged from Rosa's room and headed straight to the nearby nursing station, where Henry and I were working.

"What a wonderful reunion," she ecstatically exclaimed to us. "Rosa actually woke up for a few minutes and had a conversation with her sister. It was all really quite remarkable and very touching." Pascal's infectious, gee-shucks-isn't-it-great aura of gratitude almost alleviated the gloom around Henry and me.

"They cried a bit at first," Pascal volunteered, almost as if she were a news reporter detailing an important political summit, "but they ended up laughing about old times, about their mom and dad, and, of course, about Mama Rosa. Rosa told Maria about her friends here, especially about Yolanda. It did tire her out a bit, so we told her to sleep for now. Maria's going to spend the night with her. I really wish you two could have seen the smile on Rosa's face when Maria woke her up."

Pascal knew the anguish all the staff, especially Henry and I, had been going through in our abortive attempts to help Rosa, and her closing comment to us as she strode off the unit was typical of her watchful caring for the Spellman staff: "This reunion wouldn't have been possible if it hadn't been for you two. You guys are *wonderful!*"

We were both too emotionally exhausted to dispute Pascal's assertion.

Maria had brought from Puerto Rico their family picture from decades ago, placing it on Rosa's bedside stand, mute testimony to a time when they really were a family. The only other memento with her was Mama Rosa's favorite rosary beads, the set Mama had been clutching when she died, an event whose repercussions were now quietly concluding in room 303 at St. Clare's Hospital, almost twenty years later.

Before leaving that evening, I briefly checked in on Rosa one last time. It seemed as if her final burst of consciousness had been consumed by her emotional reunion with Maria an hour earlier. Rosa was now comatose and in no apparent pain, clasping her Mama's dark blue rosary beads in her cold, yellow hands. Her shallow breathing was tranquil—a morphine drip to facilitate her transition from life would be unnecessary. Sitting at bedside, Bible in hand, was Maria, who seemed glad to see me.

"Rosa told me she made many friends here." Maria spoke slowly, deliberately, as if she were reassuring herself more than anything. "She asked me to pray for her friend Yolanda—Sister Pascal told me about her—and I'm going to visit her in a little bit, and then light candles for both of them in the chapel. I can tell my sister has been treated very well here, and for that I thank you, Doctor." Her words were infused with a sense of grateful relief that her sister's life was ending where it was ending.

Thanking Maria for her kind words, I left for home, to ponder the day's events.

Rosa M. died in the wee hours of the next morning, at 2:20 A.M., and Yolanda J. died approximately twenty-five minutes later. According to the nurses' notes, both expirations were peaceful, without pain, bleeding, breathlessness, vomiting, or other obvious problems. The on-call medical intern was summoned to pronounce both women dead. Maria was present at Rosa's death and gathered up the family picture, the rosary beads, and the few things Rosa had accumulated during her five-week stay at St. Clare's—some cosmetics, bead costume jewelry, the New Testament Sister Pascal had given her, Thomas à Kempis's *Imitation of Christ* (a gift from Henry R.), and an unfinished picture she had been working on right before she took to bed for good. It was a painting of a tree-lined street with buildings on both sides. Three partially drawn figures, probably women, were walking hand in hand down the street, over which a rainbow was arched. Maria told the nursing supervisor that she would call the hospital later in the day for arrangements to have Rosa's body shipped to Puerto Rico for burial.

There was, of course, no one to contact about Yolanda's death, and her body was zipped up into a body bag and wheeled to the morgue. Her few possessions—an old blouse, a pair of jeans, a few socks, some underwear, bead costume jewelry, and Rosa's paintings and get-well cards—were placed in a plastic bag and thrown in a corner of the unit's medication room, to be taken later in the day to the nursing department office and, if not claimed by a family member within a month, ultimately to be pitched into the garbage. Yolanda would have a $900 "city burial," an interment without a velvet-lined coffin, a headstone, or a private memorial service.

Arriving later that morning to begin another day, the Spellman staff, including me, learned about the two almost simultaneous deaths of six hours earlier. Even before reaching the nursing station and speaking to anyone, I surmised that my two sickest patients had expired overnight. Walking onto the floor and passing Yolanda's room, I saw from the hallway that her bed was empty and covered with freshly pressed sheets. Approaching the nursing station, I glanced toward room 303 and saw on the outside door a different patient name, a Santos V., who I would soon learn was a homeless IV drug user with pneumonia admitted from the emergency room only hours after Rosa's death. Henry R., who was already at the nursing station and was reviewing Santos V.'s fragmentary emergency-room note, told me what I already knew. After briefly exchanging silent nods of regret, we began discussing Mr. V.'s problems, which, according to the ER notes, included an unwillingness to be examined by the on-call doctor.

Likewise, after similarly brief expressions of sadness, the other Spellman staff proceeded with their day, but not without a profound sense of grief that was ever unique to an AIDS unit. Rosa's hallowed journey resonated throughout the staff, a collective grief that was only manifested by forced, sympathetic smiles and knowing glances whenever Rosa's name would incidentally come up. Almost predictably, the social worker from Gift of God called the floor first thing that morning, unaware of Rosa's death, to inquire if a new discharge date had been set for her yet. Perhaps the most difficult moment came when Dr. Alma K. called me later in the day to report that she had just spoken with the Spellman medical director about "the ERCP problem," and that, pending a permanent solution, she could "maybe try" to get Rosa transferred to another hospital for the procedure. My news that the issue was moot, at least for Rosa, elicited no response.

"After the wonderful care they received here, Rosa and Yolanda have gone to an even better place," remarked Sister Pascal that morning to Henry and me, both of us already too busy with Santos V. to quarrel with her characterization of Rosa's care as "wonderful."

"You know," she persisted, with a grace we could not ignore, "sometimes our greatest gift to people like Yolanda and Rosa is just *being there,* just going in each day to say hello, saying a few kind words, when most of their lives, they haven't had many kind words spoken to them. Here were these young women who came here with next to nothing, Rosa with only the memory of her Mama Rosa's love, and Yolanda with a past no one even knew, and now"—Pascal was beaming her childlike bemusement—"and *now,* Rosa and Yolanda are together again, walking the Grand Concourse of heaven, hand in hand, with Mama Rosa."

4

THE DEATH OF STEPHEN Y.

"Stephen Y." . . .

Every name from the Reading of Names signifies a unique struggle, a unique resolution of that struggle.

LIKE ROSA M. in the early course of her illness, Stephen Y. was not an easy patient to like. But unlike Rosa, he never reformed his ways, and as he lay dying, he never reverted to the innocence of childhood, perhaps because he had never been allowed to be a child in the first place. Nonetheless, he was a patient who forced me to examine and revise my own ideas about living and dying, and his memory stays with me indelibly.

A thirty-year-old gay male, Stephen was an acquired taste, a New York original who relished argument and confrontation, who never hesitated to speak his mind, and who was already dying from AIDS when I first met him. He was also someone who taught me major lessons about respect for a patient's wishes, even when those wishes *might* seem irrational and self-destructive.

Long before he became sick, before AIDS began to terrorize the gay community of the city, Stephen had studied voice at the Juilliard School, but he was never quite able to make a career of it and had to settle for a part-time job in the Metropolitan Opera chorus, supplemented by work as a waiter at one of the dozens of white-tablecloth restaurants that fronted Lincoln Center to the east. Stephen's unrealized aspirations in music were a familiar New York story—he was but one of too many talented people competing for a paucity of jobs. Because his part-time work provided no health insurance, Stephen eventually ended up at St. Clare's when he became sick. His several stays on the Spellman service tried everyone's patience, especially his own.

As a patient, Stephen always possessed an impish, little-boy look. Short and wiry even before AIDS wasted his body further, he would initially appear beatific to the unsuspecting stranger. But whenever he would start to complain about the latest mistreatment, oversight, or indignity allegedly perpetrated on him by the hospital staff, he would angrily furrow his brow, defiantly fold his arms on his chest, and settle into a pouty, peeved persona that reminded me of the precocious five-year-old terror Calvin of the popular comic strip *Calvin and Hobbes.* But the similarity would quickly end as Stephen would continue on and embellish his complaints with a stream of colorful, choice expletives that would make a sailor blush. These withering invectives against real or imagined deficiencies in his medical care never ingratiated him to the medical and nursing staff: Stephen quickly became labeled a "difficult patient."

Stephen's rage against the imperfections of his hospital care arose from a fiery temperament that belied his slightly built frame. Within his body existed an incandescent soul that was largely not of this world, or, at least, of the world of the mundane, the pedestrian. Stephen's very being, his very reason for existing up to the very end, was music—serious or "classical" music. Although he lived for all types of classical music, opera enthralled him. Stephen was your stereotypical New York City "opera queen." Before AIDS devastated his body, he and his equally rarefied compatriots would flock almost nightly to the cheap standing-room area of the Metropolitan Opera to hear and see the world's most illustrious singers perform masterpieces with one of the best opera companies in the world. For Stephen, opera was not a hobby or a pleasant divertissement; rather, it was a consuming passion that intimated the divine and the eternal. In opera lay perfection and truth—all else was secondary, a poor imitation.

Like all opera aficionados, Stephen was merciless in his criticism of mediocrity on the stage: there could be no excuses for a sloppy performance or a passionless interpretation. For Stephen, like his idol Maria Callas, it was perfection in opera or nothing at all, and in applying his exacting standards to singers and conductors, he made no exceptions in his booing and spared no adulation in his bravos. Despite his vast knowledge of practically all aspects of opera, Stephen was never affected or patronizing toward non–opera lovers—he was an opera queen, but not an opera snob. Stephen's major problem with other people, and with life outside of the opera house, was that he transferred his ideals and high standards in music to the world beyond the Met. His demanding and haughty demeanor toward the hospital staff arose because he expected from them the same perfection and attention to detail that he expected from the performers on the opera stage, or that he expected from himself. Even as a patient, he assumed the imperious role of the impresario, the Rudolf Bing, as it were, of the hospital. But St. Clare's AIDS service was not the Met, and his AIDS story was not an *opera seria*—and therein was the source of much conflict and unhappiness.

Stephen had always been a loner, one of the many reclusive, very private gadflies who always seem to gravitate to the anonymity of the city. In one of his rare, uncharacteristic moments of candor, he told me he had had a lover many years before, and although he indicated his former lover was alive and well, he always refused to talk any more about him whenever I would try to bring it up. The few "friends" who would occasionally visit Stephen in the hospital somehow seemed more like casual acquaintances who could relate to him only through their mutual interest in Verdi, Gounod, or even Meyerbeer. They would just sit around the bedside discussing the latest opera gossip or critiquing a recent opera CD or Met production, never touching on more personal matters affecting each other. Stephen never spoke of a "best friend" and most likely did not have one. And, saddest of all, he *never* talked about his family, who lived right across the river in New Jersey, which, in Stephen's case, you would have guessed was in Tasmania. Although the social worker had informed his parents early on about his admission to St. Clare's, they never visited him or called about his condition—until, that is, the very end, when he was near death and his parents embroiled themselves in the final days of his life in a singularly monstrous way that clearly explained why Stephen had never spoken of them.

Indeed, Stephen was confronting AIDS the same way he had chosen to face most other things in his life—largely alone.

By the time I got involved in Stephen's medical care, his HIV infection was approaching the end of the line—his T-cell count was zero—and had largely been complicated by severe diarrhea, which had become more and more debilitating and demoralizing for him. During the six months prior to his final admission to the Spellman service, Stephen had had four hospitalizations, each one lasting longer than the last. Multiple evaluations during these four prior admissions failed to diagnose the cause of his worsening diarrhea. Repeated stool cultures and analyses, blood tests, bowel X rays, and even two colonoscopies with biopsies were negative. As often happens with such patients, it had to be assumed that it was the HIV itself, and not an opportunistic pathogen, that was causing the diarrhea, an assumption that was bad news for Stephen, since there was thus no specific treatment that could be given, and since symptomatic therapy with Kaopectate, Imodium, and Lomotil had not been effective in slowing the increasing torrent of diarrheal stools making his life unbearable.

In between these last hospitalizations, Stephen's growing weakness and weight loss had confined him to his small one-bedroom apartment that was only a few blocks from St. Clare's. Visiting nurses, home health aides, and homemakers frequently looked in on him, but Stephen, the "difficult patient," was often curt and demanding toward these home health workers—and seemingly ungrateful for their help. Finally, late one night, on one of his countless trips to the bathroom, he lost his footing and fell. Too weak even to get up, he lay on the floor, in an enlarging puddle of liquid stool, until the visiting nurse found him the next day, dehydrated and semiconscious. An ambulance was called, which took Stephen to St. Clare's, thus beginning his final hospitalization, an ordeal of untold suffering and—ultimately—self-discovery.

After half a day of intravenous fluids, Stephen promptly regained consciousness, and the interminable griping and yelling that defined his personality returned in high gear. Indeed, most of Stephen's complaints were entirely valid. His room's floor and bathroom were filthy, the privacy curtains were splattered with dried vomit and sputum from previous patients, his food was cold, his menu selections were routinely being ignored, the nurses did not respond quickly enough when he pushed the nurse call button, his electric bed did not work, the room's air conditioner was broken, one of his roommate's visitors stole his electric razor, and so on and so forth. These problems formed an unending litany of scathing complaints, brusquely delivered with an imperial disdain, a weary impatience that immediately made the listener defensive and equally annoyed. Consequently, the stridency of Stephen's unending demands often obscured the essential legitimacy of most of his complaints, and early on a vicious cycle was established. His grating, bitchy manner alienated many of his caregivers, who sometimes returned his rudeness and brusqueness in kind, an unfortunate response that only further infuriated Stephen.

As the first days of his hospitalization turned into weeks, his diarrhea increased exponentially in both amount and frequency, creating an unending nightmare for both Stephen and his nurses, especially as his increasing weakness made it impossible for him even to get out of bed to the bedside commode. No sooner would his soiled sheets be changed than he would have another voluminous, explosive bowel movement all over the bed. Try as they might, his nurses were never able to keep up with this fecal torrent. Stephen, as a result, spent most of his time lying in a pool of liquid stool, impatiently, almost perpetually, pushing his nurse call button, frequently yelling for the nurses at the top of his lungs, and loudly cursing the slightest tardiness in their response to his plight. The toll of this never-ending flow of feces was more than just psychic: soon his rectum became raw and irritated, with increasingly painful hemorrhoids and anal fissures.

Worsening nutrition quickly became another problem for Stephen. Despite a respectable appetite, he was never able to keep up with the fluid and nutrient loss from his diarrhea, and vitamin supplements, appetite stimulants, and even large-scale intravenous feedings failed to stem his inexorable loss of weight. Soon Stephen's muscle mass was zilch: his arms and legs became skin-covered bones, and his ribs increasingly protruded out through the thin, shiny skin covering his chest wall. Stephen was wasting away—disintegrating—before everyone's eyes.

But despite Stephen's physical deterioration, his mind remained completely intact and was largely preoccupied with his only real love in life: his music, especially the seemingly limitless trove of operas. For Stephen, as he became progressively weaker, the only reality that seemed to matter anymore was the sounds between the earphones of his portable CD player's headset. Most of his waking hours were spent listening to his considerable CD collection while intently pouring over his opera books and magazines. Dozens of classical CDs were always scattered over his bed, with many dozens more stacked on his nearby bedside stand. The usual scene in Stephen's room was quite an incongruous sight indeed: there was Stephen—usually stark naked, with his scrawny legs spread-eagle—lying or sometimes propped up in bed in a puddle of fresh stool, listening to some obscure opera on his CD player—one day Halévy, the next day Gluck, and another day Massenet—and simultaneously perusing one of his many opera tomes. Often so much stool was oozing out from around his hips that his legs and elbows would be smeared with feces, as would be his pillows and blankets. Completing this remarkable tableau would be the way he looked up from his book when he would see me come in on rounds. With his usual little frown, he would put his things aside and give me a robust tongue-lashing about the latest "atrocity"—real or imagined, although usually real—perpetrated upon him by the hospital staff. As he would let rip with his standard tirade, I sometimes felt I was really a hapless, second-rate opera tenor at a rehearsal, and he was the grand maestro bellowing criticisms and insults from the orchestra pit.

I never let Stephen's outbursts bother me. For one thing, most of his interminable complaints were fully justified, even though I could do little about the myriad deficiencies in his care. But more important, I sensed that beneath the bluster and anger was a frightened young man who had not yet come to grips with his life thus far and with his impending death. Music was Stephen's refuge, but try as he might, it was not his solace. Something was missing—call it self-awareness or inner peace. Despite his extensive knowledge of opera, his intense rage made it impossible for him to identify fully with such stoically dying characters as Puccini's Mimi or Verdi's Violetta. A unique lifetime of conflict, anger, and lost hopes was now condensed to the inner struggles, conscious or unconscious, being played out in Stephen's gaunt head as he lay in his shit-stained and shit-soaked deathbed of AIDS.

Certainly I had no idea what life issues were festering in his psyche—perhaps issues of sexuality or maybe self-esteem or breaking up with his lover. Perhaps there were family conflicts or issues about God or religion or the ultimate matter of fear of death. Stephen Y. was not a heroic figure; rather, he was the embodiment of the archetypical Everyman who faces death alone and confused. Try as he might to fantasize his predicament, to deny the terrible reality of what was happening to him, his rapidly ending life was not the neat, tragic final act of *La Bohème* or *La Traviata*.

Thus, as is often the case on the Spellman service, Stephen's real struggle was not with AIDS, but with himself. Even if he were to have a spotless hospital room, with prompt and ever-attentive nursing care and the most up-to-date medical treatment, he could still never adequately deal with his HIV disease—he would still fume and sputter on about this or that imagined slight or imperfection. Consequently, I knew from past experience that any attempts to get Stephen to focus on his HIV disease and his pending death were like putting the cart before the horse: such attempts, however well intentioned, were doomed to failure and would create only more emotional pain.

During this early part of his final hospitalization, the best I could do for Stephen was to forge a bond of honesty and trust between us. Stephen was *not* crazy. He respected candor and honesty, and to his credit, he could take well-intentioned rebuke as well as he gave it. I knew there was a fine line between being cruelly disrespectful toward him as opposed to criticizing him on his more outrageous outbursts against the ward staff. Stephen likewise understood this distinction and, although he would never admit it, appreciated being treated as an adult rather than a pitiable, whiny child. Thus, whenever he would lambaste me with a particularly unjustified tirade about his nursing care, I would jolt him back to reality by forcefully telling him he was absolutely wrong—that his unending diarrhea was *not* the nurse's fault, that his venomous remarks only alienated his nurses, and that his profound weakness rendered him totally at the mercy of his nurses. I pointed out that the nurses were only human and resented his blaming them for his problems, and that, if he was smart, he might want to treat these caregivers with a tad more respect and courtesy.

"So give the nurses a break," I would dryly conclude, "because they're overworked and don't need to hear your bitching about problems they can't help, especially when they're doing the best they can to keep you clean and happy."

My frank rebuke ended, I always knew I had successfully made my point with Stephen whenever he would remain silent and look back at me with that familiar perturbed look. To complement my no-nonsense approach toward Stephen, I would try to earn his trust by taking action on those complaints that were indeed justified—the botched-up menu request, the filthy bathroom that had not been cleaned by housekeeping for days, the occasional mice infestation in his room, the rare staff member who I sensed genuinely had it in for him. Stephen knew deep down when someone—doctor, nurse, PA, nursing aide, housekeeper—really empathized with him, and "thank you" was definitely not an unfamiliar phrase from his lips. Stephen's "thank you" always denoted respect for a job done well, much like his "bravo" at the Met.

More frustrating for me than his frequent tirades was Stephen's total denial of his condition. Perhaps his music provided him a perfect universe unto itself, but he would never really listen to my oft-repeated, bleak assessments of his situation. One memorable instance occurred several weeks into his hospitalization, when, without prior warning, he urgently requested he be allowed to leave the hospital "on pass" the next day so he could attend a recital by the Emerson String Quartet at the Metropolitan Museum of Art. He earnestly explained how he had bought tickets six months ago in anticipation of the performance's being sold-out, which it was. When I reminded him, initially patiently and calmly, that he was way too weak to stand such a trip, even for a few hours, he impatiently persisted and angrily protested that he was going anyway—that I was being unfair to him, and that it was a "very important" recital, as if I were an ignorant Philistine who just did not understand how such a "very important" musical event must simply not be missed. Exasperated for what seemed the one thousandth time with his all-pervasive lack of insight, I lost my patience and, with a bit too graphic and brutal flourish, told him to "get real"—that he did not even have enough strength to wipe his ass, let alone get out of bed to travel to a recital thirty blocks away, and that even if friends took him, how did he propose to control his massive, involuntary bowel movements during the performance?

"It wouldn't be a very pretty sight—or smell—for everyone else in the audience, Stephen," I concluded, with world-weary, give-me-a-break impatience. Angry and chastened, he fell silent and scowled at me. Never again did he ask to go out on pass. I, too, was angry, but with myself, for hurting him with the truth.

But my final struggle with Stephen centered on an issue far more important than a string-quartet recital—namely, the issue of cardiopulmonary resuscitation, CPR. As Stephen's hospital stay dragged on from weeks to months, he developed one complication of prolonged hospitalization after another—recurrent bacterial pneumonias, ever-enlarging bedsores, recurrent blood infections from infected intravenous needles, and severe anemia requiring multiple blood transfusions. Any one of these complications in a debilitated, bedridden AIDS patient could be life-threatening, especially given the profound malnutrition of a patient such as Stephen. After two dismal months of such ever-worsening problems, Stephen's situation was apparently hopeless—apparent, that is, to everyone but Stephen. I knew that if he were ever to have a cardiopulmonary arrest—that is, if he were ever to stop breathing or lose his heartbeat—his chances of being successfully resuscitated were slim. And even if CPR did miraculously bring him back from the brink, his extreme malnutrition would make it utterly impossible for him ever to get off the life-support breathing machine. He would "live" his final days or weeks as a pitiful slab of disintegrating protoplasm, almost literally pounded into bits by the high-tech machines of the intensive care unit. In such a scenario, Stephen would be "lucky" if the initial cardiopulmonary arrest knocked out his brain function, so that he would not have to suffer the torture of being conscious as his body was invaded, probed, and assaulted by the high-tech devices. Such blessings of coma from the outset of cardiopulmonary arrest could not be assured, however, and a living hell of days or even weeks of consciousness was always a possible consequence of an initially "successful" resuscitation.

From the very first time I raised this issue with him, Stephen repeatedly refused to consider signing a DNR. Characteristically, the more I would urge him to do it, the more resistant he would become, like Tosca, adamant in her spurning of the villainous suitor Scarpia. To my puzzlement and exasperation, he would never tell me why he did not want to discuss the DNR matter. "I don't want to talk about it, just resuscitate me," he would curtly reply, then go back to reading his opera books, studiously ignoring my attempts to pursue the subject further.

As the weeks wore on and his condition's downward trek continued unabated, my entreaties to Stephen about CPR and DNR issues became more urgent, primarily because I knew he could succumb at any time to one of the numerous complications afflicting him. Thinking that he did not fully realize what CPR and life-support systems entailed, I would try to describe life in intensive care on a breathing machine in lurid terms: how he would have tubes in every body orifice (nose, mouth, penis, rectum), how he would probably have his hands and legs tied down so he could not pull out these tubes, how he would be sleep-deprived and eventually psychotic from the constant noise and stimulation swirling around the intensive care unit. I emphasized how it was extremely remote that he would ever recover and get off the breathing machine, and how it would be difficult, if not impossible, to ever disconnect him from all of these tubes and let him die in peace once the threshold of CPR was crossed. Day after day, especially when the pace of his decline accelerated, I would repeatedly emphasize how he did not seem to understand what horrendous atrocities medical technology could wreak on his mind and body—that simply because medical science could keep his body "alive" for a few extra hours or days after a cardiopulmonary arrest did not make it right.

My exhortations were to no avail. The more I begged or cajoled him to become a DNR, the more steadfast he became in his refusals. He never responded to any of my horror stories about CPR and never attempted to justify his refusal to sign a DNR. It was simply an issue the two of us could not even begin to navigate. Stephen's persistent refusal to heed my warnings perplexed and angered me. At first, I thought he was ignorant about how horrendous CPR could be, how miserable life on a breathing machine in intensive care could be. I tried to theorize that his apparent pigheadedness was his way of exercising what dwindling control he still had over his body, especially since in his constricting world he could not even control his own bowel movements, let alone be independent enough to attend an Emerson

String Quartet recital. I simply could not figure it out and eventually became emotionally exhausted from my daily exhortations. Moreover, I began to feel I was becoming a bit obsessed with Stephen's code status—after all, it was *his* life, not mine, that was at issue. Finally, the day after I had twisted his arm particularly hard on this matter, I saw on my daily rounds something in Stephen's room that I will never forget. Taped on the wall directly over the head of his bed was a letter-sized piece of paper on which was printed in pencil, in weak and scrawled handwriting:

PLEASE RESUSCITATE, THANK YOU.

This was Stephen's desperate plea for life—*life at any cost,* no matter how slim the chances of recovery from CPR, no matter how agonizing being kept "alive" in intensive care might be. Stunned and dumbfounded—I had never had a patient make such a dramatic gesture for life at any cost—I thereupon resolved to stop raising the issue with him any further. His pitiful little sign over the head of his bed put the controversy to rest, at least for a time. Stephen had won the test of wills with his doctor, *as he should have.*

A little over a week after posting his plea, Stephen Y. got his wish. Late in the night, without really much warning, he became confused and severely short of breath, probably from a sudden worsening of his latest bout of bacterial pneumonia. The ward nurses STAT-paged the on-call medical intern for emergency help, and Stephen was promptly intubated—a breathing tube was rammed down his throat, into his windpipe. He had not suffered a cardiac arrest requiring a "full code," with pumping on his chest and many of the other hectic measures of CPR, but Stephen had nonetheless had a respiratory arrest, his respirations having become so burdensome that his breathing was on the verge of stopping altogether. After the emergency intubation, he was promptly transferred to intensive care, where a mechanical ventilator—a breathing machine—supplanted his weakened natural breathing. I did not learn about Stephen's crash landing until later that morning on my arrival on the ward, and I immediately headed over to the ICU to see him. The final act of Stephen's own *opera verità* had begun, in the ICU's cauldron of medical high technology.

Stephen had a private room in the ICU, and in contrast to his regular ward room, it was sterilely clean. The chamber had an antiseptic, peroxide smell to it. I reflected on the irony of how it took a respiratory arrest to get him a spotless hospital room, away from the "dump" he would call his ward room. Also unlike his previous room, his ICU room was brightly lit—in fact, almost too brightly lit, for the overhead high-intensity ceiling lights glared down onto the bed, practically blinding anyone looking directly up at them, as the patient lying flat on the bed must do. The window air conditioner was loudly blowing out gusts of frigid air, which made me chilly, even with my white coat on. Stephen's Sony clock radio, which had apparently been brought over from the ward earlier in the morning, was on the windowsill, blaring out hard-rock music over the grinding vibrations of the air conditioner's overworked, noisy compressor pump. The TV hanging from the ceiling corner opposite the bed was tuned to the Cartoon Network.

Competing with the radio and TV noise was the attendant cacophony of medical high technology: the heart monitor's staccato beeping at every one of Stephen's heartbeats, the intravenous-fluid machines quietly humming as fluids and antibiotics were pumped into his veins at ever so precise rates of infusion, and the hissing and soughing—the relentless whirring and moaning—of the breathing machine as it forced oxygen-enriched air into his lungs. Often, however, this symphony would become discordant and grating, like a bloody denouement of an eerily discordant, yet strangely rhythmic, Shostakovich opera. Various buzzing alarms would go off loudly whenever his heart monitor picked up an erratic or skipped beat, or whenever the IV pump sensed the fluid infusion rate was not keeping up to proper tempo. The rule rather than the exception, this annoying cacophony of alarms was a continual reminder that things in Stephen's ICU room were not working exactly the way they should—an auditory testament to the chasm between neatly precise medical theory and actual medical practice. Despite medical science's vaunted claims of ever-increasing knowledge—despite the impressive banks of monitors, computers, and invasive hardware that can measure a person's vital signs and organ-system functions with great precision—the inexactitude of medical science is never more evident than it is in its temple of high tech, the ICU.

As I entered his room, Stephen's two ICU nurses were there, changing his stool-soiled bedsheets. From my somewhat limited view at the doorway, he appeared awake and, like a large sack of potatoes, was rolled onto his side, held in place by one nurse while the other cleaned his bottom and legs and put down clean sheets. The nurse washing his bottom was soundly scolding him for dirtying his bed.

"You should call for a bedpan when you have to go," she lectured, never quite explaining to him how he should call for his nurses with a breathing tube in his mouth and with leather braceletlike restraints tying his hands down on the bed.

Finished with her rebuke, the nurse returned to her conversation with the other nurse about the blind date she had had the previous night and how dreadful the food was at the Italian restaurant. "I don't know which was worse, the meal or his looks," she chuckled. After the quick sheet change, the nurses firmly retied Stephen's wrists and ankles to the bed frame, sternly admonishing him not to try to pull out his breathing tube again. Their chore completed, they briskly exited, chatting about their upcoming vacations and leaving Stephen uncovered in the cold room.

Stephen lay spread-eagle on the bed, naked save for the leather straps on his wrists and ankles, as well as lamb's-wool bootees on his feet, the latter to prevent pressure sores on his heels. At least half a dozen intravenous poles were clustered around the bed, holding up not only a dozen bags of IV fluids and antibiotics but also the small electric infusion pumps that regulated the IV flow rate. Streaming from the IV poles and from nearby monitoring screens were plastic IV tubing and monitoring wires that crisscrossed Stephen's chest and abdomen in haphazard fashion. Dwarfing these smaller IV tubes and electrode wires—and almost eclipsing Stephen's tiny head and neck—were the two plastic hoses connecting him to the breathing machine. Sky blue in color and almost two inches in diameter, these two tubes were his lifelines. They snaked across his upper chest, from the tip of the breathing tube in his mouth all the way to the breathing machine, which was hissing and sighing next to his bed. The breathing machine itself looked innocuous enough, like a small portable dishwasher or washing machine. In a way, this mechanical ventilator ultimately functioned much like a household appliance. It efficiently and predictably exchanged Stephen's polluted, exhaled breaths with freshly cleaned air, as one of the connecting tubes carried out the fetid air and the other pumped in the good air.

As I had warned him, every orifice of Stephen's body was intubated: a Foley catheter drained urine from his bladder, a nasogastric tube taped to his nose emptied greenish brown gastric juices from his stomach, the endotracheal tube in his mouth was attached to the bedside breathing machine, and a small probe in his rectum continuously measured his temperature, which was undoubtedly subnormal from the chilly air blowing over his uncovered body. Intravenous catheters in his neck and both arms delivered antibiotics and fluids, while an arterial catheter in his left wrist's radial artery instantaneously measured his blood pressure. Cardiac electrode patches were pasted to his chest and were connected to the electrode wires coursing to the heart monitor over the head of the bed. Clipped onto his right index finger was a small sensor—it looked like a small clothespin—that continuously measured the oxygen level in his blood. All of these various monitoring devices displayed their important data on a set of three small TV screens perched on a shelf overhanging the head of the bed. All of Stephen's vital signs were reduced to a continuous, never-ending stream of green lines undulating and oscillating on these monitoring screens, which in turn relayed their data to the ICU nursing station in the outside hallway.

The overhead lights brightly beamed down onto the bed and onto Stephen. Despite the air-conditioning, the room felt claustrophobic. The smell of peroxide disinfectant and fresh stool—combined with the grinding of the air conditioner, the beeping alarms, and the radio and TV noise—all of these raw assaults to the senses charged the atmosphere in Stephen's ICU room with an irritable, almost repellent edge, which made me want to get the business at hand over with as soon as possible. If I felt so immediately uncomfortable in his room, I wondered, *how must Stephen feel?* Off in one of the ceiling corners, almost unnoticed and silently surveying this collective monument to medical technology, was a small video camera focused on the bed, relaying Stephen's every movement to a small TV screen at the outside nursing station—a medical Big Brother that deprived him of any remaining trace of dignity or privacy.

Seemingly entombed by all of this high-tech paraphernalia, like Aida preparing for death, Stephen's frail body appeared to play a mere supporting role to the machinery itself.

Clearing a pathway through the helter-skelter maze of IV poles, I approached the bedside. Stephen immediately recognized me and quickly furrowed his brows with the pained, peeved look that had previously greeted me whenever he was angry. I uttered some pitiful words of comfort, as if any words could mitigate the nightmare he was going through. Trying to ignore the extremity of his condition—as if this were just another routine visit on my daily rounds—I proceeded to ask him about his symptoms, such as pain, shortness of breath, and so forth. He would impatiently turn his head in the negative to every symptom I rattled off. I sensed very early on that, despite his denying the symptoms I would mention, Stephen was concerned, upset about something, but the breathing tube in his mouth, of course, made conversation impossible for him. He impatiently tried to mouth some words around the tube, but, try as I might, I could not make out what he was trying to say. Still attempting to find out what was bothering him, I ran more symptoms past him and even repeated some I had just mentioned before, in a pathetic game of medical charades. Yet he continued adamantly to deny all symptoms I would recite, becoming increasingly agitated the more I tried to ferret out what was upsetting him. *Something* was bothering him mightily, but I had no idea what it was. Was it that he was cold, I asked, covering him up with a blanket? Was the ceiling light glaring overhead too bright? Was one of the many tubes invading his body upsetting him? To these and several other guesses I proffered, he angrily shook his head no with increasing annoyance.

With frustration mounting for both of us, I stopped my questioning and went on to examine him. Although the physical examination of an ICU patient such as Stephen is easily overlooked amid the many machines and monitoring devices, his exam that first day in ICU was not much different from what it had been during the weeks before. Perturbed and impatient, as always, with my proddings and pokings, he was a hapless bag of skin and bones—yet the indomitable spirit persevered, as I continued to spar with him.

By the end of my brief exam, Stephen was *profoundly* upset with me. With great effort and animation, he was desperately trying to mouth something to me around the large breathing tube in his mouth. What important, obviously urgent thing did he want to say? Was it a symptom I had not inquired about? Was it a plea to be disconnected from the breathing machine—to die in peace? Or was it something mundane, like a request for a bedpan? When, in desperation, I began repeating the laundry list of symptoms I had already gone over, he closed his eyes tightly and shook his head violently from side to side, as if he were being tortured. Afraid he would dislodge his breathing tube or injure himself some other way with his tantrum, I decided it was best to give up for the time being and leave Stephen to rest. Apologizing for being unable to understand him, I started to turn away, at which point he forcefully jerked and waved his strapped-down right hand in a writing gesture. He was desperately trying to signal that he wanted to write something to me!

Feeling like a total idiot—after all, Stephen could still communicate by writing even though he could not talk—I quickly got a writing pad and pencil from the nursing station, untied his right hand, and held the pad up as he attempted to write what was so important to him. So weak that he could barely hold the pencil, he had to make several tries before he was able to scratch out two words that shamed me and made me feel even more of an insensitive jerk:

MY MUSIC

Overwhelmed by my thoughtlessness—I had been focusing only on his physical condition and had completely ignored how someone earlier in the day had tuned the TV to the Cartoon Network and his radio to a hard-rock station—I quickly turned off the TV and walked over to his radio on the windowsill and turned it back to WQXR-FM, his usual classical music station. A Bach trio sonata immediately filled the room with unspeakable clarity and authority, and a perturbed Stephen grudgingly nodded his thanks to me. As the music pierced through the ugly sounds of the ICU, the beauty and genius of Bach somehow seemed an irrefutable indictment of the obscene violation of Stephen's humanity. Somehow, that simple trio sonata, which Bach had tossed off in a few hours, conveyed more truth and beauty than all the technology represented by the machines and monitors surrounding Stephen. Returning to his bedside, I held his hand and apologized for allowing all the ICU equipment on his body to get in the way of my looking after what was really important—his music. By now exhausted, he weakly signaled his satisfaction and drifted off to sleep.

Leaving Stephen to his dreams, I went to the ICU nursing station and wrote in his chart the orders, "Patient's radio is to ALWAYS be tuned to 96.3 FM, WQXR, and is NEVER to be changed."

Not surprisingly, Stephen's ensuing days in intensive care only became more dismal. Although I was still his attending physician, responsible for all major decisions in his care, the minute-by-minute critical-care decisions had to come from the hospital physicians staffing the ICU, and I consulted with them daily about his case. His pneumonia, which was the immediate cause of his being put on life-support systems, steadily worsened. His ever-weakening lungs became increasingly inefficient in oxygenating his blood, and the prospects of getting him off the breathing machine grew increasingly bleaker. The inevitable complications of being hooked up to a breathing machine set in. His face and neck developed disfiguring puffiness and swelling from the machine's forcefully pushing air into the body's soft tissues around his windpipe. His head and upper chest area became bloated, and his eyes bugged out unnaturally. Several IV sites in his hands and arms became swollen and infected. His already fragile skin started to erode away into pressure sores on his elbows, lower back, and buttocks, especially since the prolonged bed rest and leather restraints kept him from moving and repositioning himself in bed.

But the greatest, and saddest, casualty of Stephen's ICU stay was the psychic toll on him. He quickly became sleep-deprived from the incessant visits in his room by the ceaseless army of ICU staff assigned to his care. His nurses made multiple visits to change his bedsheets, to suction out lung secretions from his breathing tube, to hang IV fluids and antibiotics, to change the dressings on his IV sites, to empty his urine bag, to make sure the monitors were properly attached to his body, to turn and reposition him in bed, and to clean his bedsores. No sooner would the nurses leave than he would be disturbed by the respiratory therapist checking his breathing machine, or by the phlebotomist drawing his blood for twice-a-day testing, or by the housekeeper, at long last cleaning his room. He would be awakened and prodded by the X-ray technician taking his daily portable chest X ray, or by the EKG technician hooking him up for his daily EKG, or by his physicians, the psychiatrist, the social worker, or the pastoral care staff. Stephen, like most critically ill ICU patients, was lucky if he got even an uninterrupted hour or so of quiet time. Soon he lost any sense of time, any sense of day and night, and as often happens in such patients, he would intermittently become confused, disoriented, and psychotic.

Patients who have fortunately recovered from prolonged ICU stays for life-threatening illnesses often later recall having had frequent nightmares and hallucinations while in ICU, sometimes involving macabre visions of being at their own autopsy or funeral service. Given the incredible emphasis ICU care often places on *things* rather than people, some patients develop a feeling of acute depersonalization. That is, they feel literally as if they are living outside of their bodies, and a common hallucination has the patient viewing himself in the ICU bed from an imagined vantage point at the foot of the bed or from over the bed. Sometimes ICU patients become obsessed with a particular technical aspect of their illness, especially an aspect that they have learned from their doctors and nurses to be crucial for their survival. For example, a patient with a serious heart attack might fixate on his heart rate or blood pressure, or a patient with a bleeding ulcer will obsess over his blood count. Often these obsessions occur to the total exclusion of everything else around the patient, including friends and family. Fortunately, all of these psychological traumas of ICU existence can resolve completely, *if* the patient recovers from the serious illness. However, by the time Stephen reached ICU, his chances of recovery were so infinitesimally small as to be practically nonexistent.

Given these well-known psychological reactions most seriously ill patients have to the many stresses of ICU life, it was not surprising that, after that first day in ICU, Stephen would only intermittently recognize me when I made rounds. When on subsequent visits I would try to communicate with him, his responses—written or by gesture—were often inappropriate or confusing. Sometimes when I would ask him a question, he would stare at me blankly and not attempt to reply, whereas other times he would simply appear terrified—whether at me or at some hallucination, I had no idea. These reactions of terror were particularly pathetic. With the whites of his eyes bugging out in full stare, he appeared like a dumb animal caught in a trap, helplessly dangling on the end of the breathing tube, like an animal snared by its mouth or snout. Other times, although less and less frequently, Stephen would seem lucid and rational in his responses. I knew he was back to normal whenever he would appear annoyed and perturbed with my questions, but the characteristic spark of defiance and independence flashed less and less often and with diminished intensity. Most sadly, he never listened to his CDs anymore, and his opera books and magazines remained piled on a chair in a corner, closed and untouched. But his radio remained permanently tuned, as prescribed, to the comforting melodies on WQXR-FM.

On both the fourth and fifth days of Stephen's ICU stay, there came a development that I had feared. On three occasions, while the nurses were not watching, he somehow managed to break loose from his wrist restraints and yanked the breathing tube out of his mouth, only to rapidly develop labored breathing and impending respiratory failure. The doctor in the ICU was quickly summoned, and Stephen was reintubated and placed back on the breathing machine, with resultant relief of his breathlessness. I dreaded such a development because I did not know what to make of it, and because it unavoidably raised the issue of DNR once again. Was Stephen's pulling out the breathing tube simply because of his confusion—that is, was he just reflexively trying to remove this irritating tube from his throat? Or, much more significant, was he trying to tell everyone that he wanted off the breathing machine, that he had had enough of life at all costs and wanted now to die in peace? The previous weeks of trying to convince Stephen to opt for DNR status had made it clear that he had wanted full CPR and life support. His plaintive sign, PLEASE RESUSCITATE, THANK YOU, was still fresh in my memory. And, as I had repeatedly pounded away at him for weeks, once the line of CPR is crossed and a patient is put on a breathing machine, it is no easy matter to step back and undo what has been done. Disconnecting a breathing machine—"pulling the plug"—can be next to impossible in New York State, especially if the patient is too confused to make his wishes known and if there is no health-care proxy previously appointed by the patient to make such decisions. The late evening of Stephen's fifth ICU day he pulled out the breathing tube for a third time (twice on that day alone), and was promptly reintubated.

What, if anything, should I do? I agonized to myself.

At that point in Stephen's care, I had two choices, neither of which I liked. I could conveniently ignore his repeatedly extubating himself and rationalize that he was by then too confused to know what he was doing. He had previously made his desire for aggressive life support unequivocally clear, despite my many attempts to persuade him otherwise. Such a choice for maintaining the status quo—a fatalistic, the-die-is-already-cast approach—would be easy and quite defensible, but it would mean continuing the all-out, relentless ICU care Stephen was already receiving. The problem, of course, with this first option was that it ignored the terrible possibility that maybe, *just maybe,* Stephen had changed his mind and now wanted to end his suffering—that his pulling out his breathing tube was his mute plea for an end to living at all costs. To ignore this latter possibility would condemn Stephen to terrible suffering until he finally died, a thought that I found repellent.

My second possible response to Stephen's self-extubation was equally unpalatable: namely, once again resurrecting the entire issue of DNR either with Stephen or, if they could be contacted, with his thus far uninvolved family. This second choice would probably entail a lot of effort, emotional stress, and hassles for me, for Stephen, and for his family. Yet, because this second option might possibly alleviate Stephen's suffering, I decided to make one final try at getting Stephen made a DNR. It was going to be a monumental project, far more daunting than my earlier, spectacularly unsuccessful effort with Stephen on this matter.

To assist Stephen, and me, with this difficult matter, I needed the help of Spellman's psychiatrist, social worker, and pastoral care staff, all of whom had been involved with Stephen's care from the very start. Like me, these colleagues wanted to help Stephen live—and die—with AIDS on his own terms, whatever they might be and however they might change with time and circumstances. The psychiatrist could help me determine whether Stephen was mentally competent to make decisions about his care, an important but vexsome task, given his increasing spells of ICU-related confusion. Stephen's social worker could again try to locate his parents and coordinate efforts to get someone to be his health-care proxy, so that medical decisions could be made if and when he was no longer conscious or competent to make decisions himself. And pastoral care would continue to provide needed emotional support to Stephen, to his family (if they could be found), and perhaps equally important, to the other Spellman staff caring for Stephen and facing the daily stresses of his downward spiral.

However, just as I had feared, reapproaching Stephen about stopping life support systems was frustrating. Realizing my raising this delicate subject could anger or frighten him, I tried to sound less strident, less persistent in my discussions with him. To my initial gratification, Stephen did not automatically shut me out when I started talking about how he could still opt to be a DNR, if he wished. But his overall reactions were consistently inconsistent. One moment he seemed to understand fully what I was talking about, and then, without warning, the next moment his mind drifted off into outer space, with glazed eyes and catatonic silence. Sometimes Stephen appeared lucid and rational, only to lapse suddenly into periods of confusion, which would then usually deteriorate into one of his terror episodes. As if in a seizure, his face would remain frozen in horror, as if he were staring into a private hell, and my attempts to reorient him back into reality would only agitate him more. Other times, perfectly appropriate responses by Stephen to my comments alternated with his asking me nonsensical, contradictory questions.

Once, when I thought he really understood my point that he could have the breathing tube in his mouth removed once he signed a DNR, he wrote back to me, "When are we going to my funeral?" My frustration was shared by the other Spellman staff trying to help Stephen with this matter. The psychiatrist could not be certain whether Stephen truly grasped the issues of his medical care, and Sister Pascal was likewise never sure how well she was able to reach Stephen about anything, let alone DNR questions. But most astounding to everyone who was pressing Stephen about ending life support was the fact that *during those brief but apparently lucid moments when he seemed to understand what we were talking about, he would always refuse to terminate such life support.*

It was incredible to me that someone so moribund and obviously miserable—someone with both feet solidly in the grave—would not want to say "Enough!" and die peacefully, easing into death with the benefit of the tons of morphine and tranquilizers I had always assured Stephen I would load him up with. From my perspective, Stephen's condition was so hopeless that it would almost seem criminal to prolong his suffering. Yet, remarkably, whenever the foggy mists of his confusion and terrors dissipated and a Spellman staff member or I felt we were talking with the *real* Stephen Y., he would always be resolute. His plea of several weeks earlier—"Please resuscitate, thank you"—remained operative. Stephen's psychiatrist, the social worker, and pastoral care staff agreed with me. After several days of trying to determine if the hell of the ICU had changed Stephen's mind about holding on to his life at all costs, we had to conclude that his pulling out his breathing tube had to be attributed to his intermittent confusion and not to any attempts—subconscious or otherwise—to stop life support. Amazingly, Stephen had not given up on living.

At this point in Stephen's arduous saga, I was emotionally exhausted—and proverbially, only God knew what Stephen himself was feeling. The other Spellman staff involved with his care were likewise numbed and dumbfounded. Three and one-half months had elapsed since the ambulance had brought a dehydrated and malnourished Stephen Y. to St. Clare's for the last time, and he was miraculously alive, but barely. I felt guilty about having pushed the DNR matter too hard before he ended up in ICU, but the second go-round—when he was finally on life-support systems—was dreadful. Try as I might to be caring about his suffering, I feared he might think I was asking him to sign his own death warrant, in a way, to commit suicide, since by that point he was dependent upon the breathing machine. I rationalized to myself that my motive in resurrecting the DNR issue was good: to end his horrendous suffering, which I felt was inhumane to prolong, since death was unquestionably imminent. But focusing on his life-support systems, on "pulling the plug" on his breathing machine, perhaps made it look to him as if everyone were impatient for him to die.

Meanwhile, by the time the other Spellman staff and I had finally given up on persuading Stephen to opt for DNR status in the ICU, his condition had worsened further. A highly virulent fungus infection originating from one of his intravenous lines spread through his bloodstream and damaged one of his heart valves, necessitating the addition of a potentially toxic antifungal antibiotic to the three antibacterial antibiotics he was already getting.

As often happens in critically ill patients, these antibiotics began to injure Stephen's kidneys, and his urine output started to dwindle to nothing. A kidney specialist was called in, and the possibility of dialysis—hooking him up to a kidney machine—was seriously mulled over, the consultant's decision being to "temporize"—medicalese for "watchful waiting"—rather than immediately dialyzing him. By fiddling around with his intravenous fluids and changing his antibiotic dosages, the ICU doctors and I were able to arrest his kidney failure for the time being, but these measures would only be stopgap, like holding off a forest fire with a small fire extinguisher.

Stephen now slept most of the time but was still arousable and, remarkably, was still able to communicate with me by scribbling notes. But gone was the fierce stubbornness I had come to respect. In its place appeared an overwhelming fear and pathetic helplessness that instantly transfixed his bloated little face whenever I would awaken him from sleep. Perhaps most tragic was how his few "conversations" with me and other staff now centered exclusively on his concerns about the breathing machine. His bedside notepad was strewn with worried admonitions and inquiries, such as, "Is the machine OK?" or "Don't take it [the breathing machine] off" or "No DNR." My repeated reassurances about leaving the machine hooked up to him did not ease his mind on the matter. While often typical for such a critically ill patient, his new fixation with his life-support machines filled me with shame and regret, since I feared, rightly or wrongly, my previous efforts to get him to acquiesce to DNR had warped his mind with this issue. It seemed that Stephen's unique lifetime of hopes, memories, and experiences had been constricted and focused down to a breathing tube and the machine it was attached to. Yet his bedside radio still remained tuned to classical music day and night, connecting him, consciously or unconsciously, with his earlier life.

Stephen's repeated refusals to let go—to acquiesce to an end to his suffering—touched deep anxieties and uncertainties I myself had had about dying. Stephen's situation epitomized the old "to be or not to be" conundrum. It was impossible for me to contemplate his worsening condition day after day without putting myself in his place: How would I myself react under such circumstances? Perhaps my peace of mind would have been less vulnerable if I had convinced myself that Stephen's suffering—or, for that matter, the suffering of all of my other AIDS patients—did not have any relevance for me personally. Medical school training typically conditions doctors against becoming too emotionally involved with their patients. Indeed, if doctors were to personalize too strongly their patients' sufferings, they would either become gods or go insane in the face of the pain and death they face every day. To protect themselves, and to maintain clear clinical judgment, doctors insulate their emotional reactions from their patients' sufferings.

But this protective tradition is good only up to a point. Problems arise when doctors go overboard in this self-protection, to the point of denying, consciously or unconsciously, that the suffering and dying they see every day have any significance for their personal lives. Such daily denial, as well as the control and power they still wield over their patients, is why most doctors are ironically least prepared for their own eventual illnesses and deaths. Therein lies the fine line a good doctor must walk when caring for a terminally ill patient: not to get emotionally overwrought with the patient's suffering, but also not to harden that emotional detachment into self-denial of death's role in the doctor's own life. *Such appreciation of the personal significance of death, in addition to the inevitable corollary of the wonderful uniqueness of every person's life, is what should make being a physician a great privilege.* These insights into life and death are definitely not limited to medicine and can be arrived at through a variety of disciplines—the arts, music, philosophy, history, science, or simply living the examined life espoused by the ancients. But by constant trafficking with suffering, a doctor should allow these issues to resonate harmoniously in his or her soul, the resultant appreciation of life thereby comforting those patients facing their own mortality. Unfortunately, being all too human, many doctors avoid these weighty questions altogether, especially when confronted with overwhelming, unprecedented cases like Stephen Y.'s. Too often, doctors let money, pride of scientific knowledge, or life's usual daily worries distract them from assimilating the lessons of their patients' suffering into their own lives.

Thus, Stephen Y.'s virtually unimaginable hospital course—his unbelievable insistence on life at all cost—engendered intense introspection. How could someone still want to "live" under such horrendous conditions when there was no hope whatsoever—when death was so palpably near? What exactly had been going through Stephen's head during this ordeal, and, more important, what would be going through *my* head if I were in his place?

A valuable insight into these questions came to me from a junior colleague, a PA student who had been following me on my hospital rounds during the several weeks Stephen had been languishing in ICU. Sensing my turmoil and frustration with Stephen's insistence on life at all costs, this perceptive student suggested that perhaps Stephen did not want to die until he settled, consciously or unconsciously, some important life issues.

"Who are we to say when it's time for Stephen to die, when he may have some unfinished business to deal with," chided my PA student.

As trite as it sounded on first hearing, this observation had an unmistakable ring of truth. Indeed, as with most things in life, the answers to weighty existential questions are usually simple and direct, and such might have been the case with Stephen Y. My student cited Leo Tolstoy's *The Death of Ivan Ilych* as the classic story of the previously superficial, nonintrospective person facing the terror of impending death, desperately realizing, in Ivan Ilych's words, "What if my entire life, my entire conscious life, simply was not the real thing? . . ."

> It occurred to him that what had seemed utterly inconceivable before—that he had not lived the kind of life he should have—might in fact be true. It occurred to him that those scarcely perceptible impulses of his to protest what people of high rank considered good, vague impulses which he had always suppressed, might have been precisely what mattered, and all the rest had not been the real thing. His official duties, his manner of life, his family, the values adhered to by people in society and in his profession—all these might not have been the real thing. He tried to come up with a defense of these things and suddenly became aware of the insubstantiality of them all. And there was nothing left to defend.

Ivan Ilych had led a safe, successful, and self-satisfied life as a nineteenth-century Russian Imperial High Court judge, a man who never regarded death as anything to do with him. It always happened to others, but never to him. But when, in the prime of life, he is faced with the inevitability of a slow and terrible death, Ivan Ilych's previous disbelief about death is transformed into terror and hopelessness. With wonderful genius, Tolstoy tells the gripping psychological drama of a lost soul groping for answers:

> "But if that is the case," he asked himself, "and I am taking leave of life with the awareness that I squandered all I was given and have no possibility of rectifying matters—what then?" He lay on his back and began to review his whole life in an entirely different light.

Finally—and here was, I believe, the crux of the issue for Stephen Y.—Ivan Ilych's salvation came with the transcendent realization that *it was never too late to put one's life in order.*

> "Yes, all of it [Ivan's life up to that point] was simply not the real thing. But no matter. I can still make it the real thing—I can. But what *is* the real thing?"

This remarkable insight came only one hour before Ivan Ilych's death, but it was enough time for him to discover "the real thing" and die at peace, as a complete human being and not the empty shell he had been for most of his life.

Was it possible that, like Ivan Ilych, Stephen Y., too, was struggling with what the real thing of his life was? Although his body had been disintegrating under the ravages of AIDS, his soul had very possibly been growing stronger as he wrestled with his life's past and present demons. Ultimately for Stephen, the final struggle was not with the flesh but with the human spirit's trying to make sense of it all. Viewed in this light, Stephen's ordeal seemed a necessary journey for a troubled soul. The image of Stephen in extremis in the intensive care unit was no longer an obscene, cruel joke for me. Rather, it was a picture of a person determined not to die with an unfinished life.

These sobering realizations lifted a great weight of concern and responsibility off my shoulders. As Stephen's physician, I now saw my job as that of a benign, passive facilitator to help him live—and die—on his own terms, in his own time. I felt I was no longer an accomplice to an inhumane prolongation of human suffering. I was now at peace with the Stephen Y. case, and my goal was for Stephen to be at peace as well, even if that peace arrived, like Ivan Ilych's, just moments before his final breath.

But Stephen's final peace had one more obstacle, perhaps the most formidable one of his entire life—his parents. Indeed, as horrendous as it was, perhaps it was his parents' arrival that Stephen was waiting for, so that he could bring final resolution to his struggle for the "real thing" in his life.

Throughout Stephen's final three-and-one-half-month journey at St. Clare's, only his older brother contacted me about his condition, and then only sporadically. Although he never visited Stephen, he did always promise me he would ask their parents in Elizabeth, New Jersey, to call me for a condition report, but he cautioned that they had been estranged from Stephen for some time and might not want to talk. Mr. and Mrs. Y. never called, and when I twice telephoned what was listed as their number in New Jersey, I heard a weird recording that sounded like a dial-a-prayer hybrid for a right-wing Pentecostal Catholic organization. A shrill male voice was screaming tirades against "baby-killer abortionists, and God-hating queers," with apocalyptic promises that the Blessed Virgin Mary would soon return to earth to "cleanse the world of such vermin." The recording left no opportunity for a message, and I assumed—hoped—I had reached a wrong number.

Finally, after Stephen's transfer to intensive care and when the matter of "pulling the plug" came up, a Spellman social worker was somehow able to reach his parents and inform them about their son's bleak condition. To everyone's amazement, Stephen's mother and father showed up unexpectedly a week later, just after the second go-round with Stephen about DNR had concluded with the decision to drop the matter once and for all.

As soon as Stephen's parents arrived in the ICU for their first visit, the nurses there—per my prior instructions—immediately contacted me and had them wait for me in the ICU's waiting room, located next to the ICU. I wanted to talk with them before letting them see Stephen. I wanted to prepare them for their son's terminal condition, especially since they had not seen him for quite a while. When I walked into the waiting room, I instantly realized why Stephen had never spoken of them before.

Stephen's father looked a lot as Stephen probably once did—short, wiry, and with eyes full of intensity—but even a cursory glance convinced me that Mr. Y. was a total fanatic, a religious lunatic of the first order. A fidgety, nervous man in his late fifties, he thoroughly dressed the role of the God freak: white socks, black sneakers with orange laces, long-sleeved plaid shirt (buttoned tightly at the top), white plastic pen guard with a half dozen pens in his shirt pocket, thick, black-framed eyeglasses broken in several places and patched up with dirty white tape, and not quite long enough work pants held up by red suspenders on which were pinned a dozen or so anti-abortion, antigay, anticommunist, "pro-family," and "pro-life" buttons. Similar buttons were pinned on the sides of the Yankees baseball cap he wore, and around his neck were several votive amulets of the Sacred Heart of Jesus and the Blessed Virgin Mary. He was carrying a worn Bible, a supply of what appeared to be religious tracts, and a tattered brown briefcase with anti-abortion and antigay stickers plastered all over it. I did not even want to contemplate what might be inside it. Complementing his eclectic ensemble, Mr. Y. had a bizarre, intense stare, and I immediately sensed I was dealing with a man on the edge of reality, if not already beyond it. Mrs. Y., on the other hand, was an apparent nonentity, a mousy, emotionless woman who deferred to her husband and never said anything to anyone. Her husband did the thinking and talking for both of them. Only real life—especially real life, New York style—could conjure up such a pair of religious cranks, I thought to myself, but they seemed harmless enough, at least on first meeting.

Trying hard not to gawk at Mr. Y.'s attire, I politely greeted both of Stephen's parents and introduced myself. As I casually took in their incredible appearance, I carefully reviewed with them the torturous course of Stephen's illness and the extremity of his present condition, emphasizing the inevitability of his dying soon. I warned them that their son would look terrible, trying to prepare them for the shock of his skeletal body, the numerous tubes invading it, and the general bleakness of his condition. I explained how he had consistently wanted heroic, life-supporting measures to be taken and how he had declined my most recent offers to disconnect him from the breathing machine, on which he was now totally dependent. My mention of this last item—Stephen's steadfast refusal to sign a DNR—especially caught Mr. Y.'s attention, and I saw a strange gleam briefly light up his eyes when I spoke of the option of disconnecting the breathing machine. I told them I was sorry I could not offer them any hope for Stephen's recovery. Comfort measures and, most important, emotional support and love were what Stephen now needed, I emphasized, in a subtle appeal to their parental instincts. Finally, I thanked them for coming and said I hoped they could help ease their son's suffering and loneliness.

Except for Mr. Y.'s brief, disquieting glimmer of animation at my mention of possibly disconnecting Stephen's life support, his parents listened to me impassively, without any emotion or response whatsoever. Vaguely uneasy at their silence and lack of questions, I led them to Stephen's bedside and, gently awakening him, told him his parents were there. Other than perhaps appearing somewhat confused, he did not react to my news. I waited for only a moment, to be sure he was fully awake, and then, out of respect for their privacy, left the three of them alone for their reunion, returning to my ward.

About ten minutes later, I received a STAT page to return to the ICU immediately. As I hurried back to Stephen's room, I wondered what could be wrong. Could it be a seizure, perhaps unstable vital signs, perhaps even a cardiac arrest? Rushing into his room, I half-expected to see the nurses pumping on his bony chest in a desperate attempt to revive him, but instead I saw a far more revolting scene, a scene that I will forever remember.

Standing right up against the bed rails and malevolently hovering over Stephen's near-skeletal body was his father, waving his tattered Bible with his right hand, violently shaking his left index finger in Stephen's terrified face, and loudly ranting over and over, *"You are doing reparations for your sins! You must offer your suffering as a sacrifice to the Blessed Virgin, or you will surely die in hell!"*

Mr. Y.'s face was transfixed with raw rage, and Stephen's was frozen in wide-eyed horror at his father's howlings. Standing back in a corner was a stone-faced Mrs. Y., quietly fondling her rosary beads and hanging on her husband's every word. At that same moment, Father Jack and Sister Pascal entered the room just behind me, apparently summoned STAT by the ICU nurses, just as I had been. Beholding this incredible spectacle with open-mouthed amazement—after all, this was supposed to be an ICU room, not a Pentecostal revival tent—the three of us were momentarily stunned by this macabre tableau. Yet Father Jack sized up in an instant what was happening and moved into action.

Loudly cursing Mr. Y., Jack lurched toward him and forcibly pulled him away from the bedside. *"What the hell are you doing?"* Jack yelled at Mr. Y, who, shocked at the ferocity of the priest's pulling, quickly recoiled in fear. Indeed, Jack was at least twice the size of Mr. Y. and now defiantly stood between Stephen and his father, fists clenched and ready to punch out the Bible-thumping zombie if he made any move toward the bedside.

Quickly regaining my bearings, I ordered Mr. and Mrs. Y. out of the room at once, and when they tried to protest their expulsion, I asked Pascal to immediately call hospital security officers. This threat—as well as a livid, sputtering Father Jack commanding them *"Get the hell out of here before I kick the shit out of both of you"*—convinced Stephen's parents to exit the room without further protest.

Closing the door behind his parents, I returned to Stephen's bedside. He was trembling. He stared blankly into space, apparently too traumatized to respond to my inquiries about his condition. Letting Pascal and Father Jack look after him, I departed to hunt down his parents, my mind still reeling from the grotesque abuse I had just seen them perpetrate on helpless Stephen. My first impulse was angrily to banish them from the hospital, but I realized we were possibly dealing here with psychotics. Thus, a detached, professional demeanor was probably best.

Mr. and Mrs. Y. were huddled back in the ICU waiting room. She was still reciting the rosary, and he was now intently reading his Bible. As soon as I walked in, Mr. Y. leapt to his feet and started again to protest their treatment, but I quickly cut him off and firmly announced my intense displeasure, my eyes glaring and voice dripping with solemn, icy condescension. Their threatening behavior toward Stephen, I warned, was "totally unacceptable and will not be tolerated." They would be permitted to visit Stephen only under the supervision of either Pascal or Father Jack.

Mr. Y. seemed shocked at my indignation and, in a voice more whining than threatening, countered that what they said to their son was "our own business and no one else's," adding that they would sue and go to the newspapers if access to Stephen were limited.

"I'll even go to the cardinal if we have to and tell him what a *den of iniquity* he's running here," Mr. Y. concluded, apparently believing His Eminence really wanted to meet with yet another religious fanatic.

Having long ago learned that such threats are best neutralized by inviting them rather than quaking at them, I told Mr. Y. to go ahead and sue and go to the newspapers and the archbishop—but I coolly reminded him that a priest and nun would back up my version of things.

"Besides, Father Jack is sometimes O'Connor's personal confessor, so I doubt if you'd be wise to get the cardinal involved," I countered shamelessly, baiting them with a lie so patently outrageous—and, in Jack's case, *ironically* outrageous—that I knew they would both be dumb enough to believe its plausibility. Somehow the entire situation, as tragic and scary as it was, seemed so fantastically loony to me that it made perfect sense to counter their insanity with the equally bizarre idea that Father Jack would be His Eminence's personal confessor.

Quickly frustrated by my refusal to back down, Mr. Y. tried to become conciliatory. Their zeal, he solemnly intoned, was prompted by their recent visions of the Blessed Virgin, who told them their son was to be miraculously cured—"the world's first AIDS cure by the Blessed Virgin," he reverently emphasized—if Stephen were "to repent of his sins and the abomination of homosexuality." I let these latter remarks pass without comment, although I secretly marveled how pristinely medieval Mr. and Mrs. Y.'s minds were. I later learned from Sister Pascal that Stephen's parents were fringe members of the Church of Our Lady of Newark, where a fanatical cult had recently been preaching about miraculous appearances of the Virgin in the very neighborhood of the church. The local bishop had been trying to squelch such activities, but with limited success.

Realizing it would probably be easier, and less contentious, to let these nuts see Stephen under supervision than to try to ban them outright, I aloofly reiterated the conditions of their future visits with their son—namely, that they leave their fire-and-brimstone rhetoric at the ICU door and not talk to their dying son about hell. "I will not tolerate revival meetings in the ICU, especially when Stephen is a captive audience," I sarcastically observed, adding that Pascal or Jack would monitor their behavior. Mr. and Mrs. Y. reluctantly agreed, but Mr. Y. insisted he had the right to read to Stephen any Bible passage he wanted.

"This is America, Doctor, and this is a Catholic hospital, which means God's word is not to be denied," he piously observed. "Besides, I don't know how a priest or nun could object to me reading Stephen a Bible passage or two." Mrs. Y. was adding her silent nods of assent in the background. Deciding I was not going to discuss theology with these two, I ignored Mr. Y.'s last remarks and abruptly concluded my confrontation with them by noting that they could revisit Stephen only if he gave his okay.

Later that morning Stephen indicated to me he wanted to see his parents again, despite the harrowing religious sideshow they had put on earlier. I had long since stopped trying to predict what his responses to anything would be. As a result of my fulminations, and Sister Pascal's ever-watchful presence, Mr. and Mrs. Y.'s subsequent bedside visits over the next few days were comparatively benign, but still troublesome and bizarre.

The strange glimmer in Mr. Y.'s eyes at my earlier mention of Stephen's breathing machine had not been just a nervous tic. Rather, my raising this matter provided this crazy little man an evangelical hammer to pound away at Stephen time and time again. For, although the issue of DNR—"pulling the plug"—had already been put to rest before their arrival on the scene, his father seized upon this nonissue to admonish Stephen repeatedly that signing a DNR would be "committing suicide, a mortal sin." In between her silent rosary recitations, Mrs. Y., who otherwise never spoke, exhorted Stephen to be "worthy of the great miracle our Blessed Lady plans for you." During these bedside vigils by his parents there were the obligatory Bible readings, mostly passages dealing with sin, especially the "abomination of homosexuality" as proscribed by St. Paul and the writer of Leviticus. Never did Stephen hear of love or redemption from his father's Bible selections, and never did his parents talk with him of family news or relive happier times past. A few days later, when I was visiting Stephen alone on rounds, I noticed that someone—I really did not have to guess who—had dialed his radio to an AM religious station, which at the time was loudly broadcasting a radio evangelist's particularly rabid tirade against "secular humanists" and abortionists. I immediately turned the radio back to WQXR-FM and unscrewed the tuning knob so it could never be changed again. As extra insurance, I exploited Mr. and Mrs. Y.'s ignorant obsession with his life-support systems by attaching to the radio a small note: "Do not touch radio knobs or breathing machine might malfunction." It remained tuned thereafter to WQXR.

These visits by Stephen's parents soon became the talk of the Spellman service, and many of the physicians, PAs, and nurses would idly wander into the ICU to catch a passing glimpse of the strangely acting, weirdly attired parents of Stephen Y. Father Jack refused to chaperone their stays in Stephen's room, primarily because he was afraid his temper would get him into trouble. Mr. Y.'s gospel of sin and retribution deeply offended Jack's vision of the healing power of the Gospel. Thus, Pascal had the chore of watching Stephen's parents torment their son with their pious claptrap. On a few occasions, when Mr. Y.'s lurid accounts of Old Testament punishments became too strident, she had to intercede with a slightly impatient clearing of her throat. When that polite warning was too subtle for Mr. Y., she would interrupt his religious rantings with a cheerful but forceful rejoinder, "Yes, but I particularly like the passage where Jesus said . . . ," and here she would abruptly take over the conversation and recite to Stephen a brief story or Scripture section about God's love and forgiveness. Although short in stature and gentle in temperament, Pascal also had a New York street-tough side to her that she could use to great advantage when dealing with troublesome people such as Mr. Y.—a don't-mess-with-me approach that somehow melded perfectly with her incandescent warmth. Mr. and Mrs. Y. were no match for Pascal's mélange of love and hard-assed attitude, as she tried to protect Stephen from their more egregious ravings.

As if to confirm his status as a first-degree loon, Mr. Y. taped on the walls of Stephen's ICU room six or seven glossy eight-by-ten-inch photos of what he claimed were his visions of the Blessed Virgin Mary. To another observer, they were simply photos of a church's front yard, and the "visions" were just obscure shadows or indistinct sun glare. Immensely proud of these "visions," which could miraculously be recorded by a camera, Mr. Y. showed his wondrous souvenirs to whomever he could corner, including nurses, housekeeping staff, even visitors of other ICU patients. Whenever Stephen would be conscious, Mr. Y. would hold one of these pictures up to his face, as if it were a sacred icon for him to meditate upon. A perplexed, worried-looking Stephen would knit his brows in intense concentration as his father would excitedly point out for him which shadows were really the vision of the Virgin. Stephen's ICU room soon had the trappings of a religious grotto, a mini-shrine to the Virgin. As if Fátima were being replayed in Hell's Kitchen, several small statues of her were on both the bedside table and the windowsill, and a dozen or so rosary beads and other sacred amulets were strung about the IV poles and bed rails. On top of the breathing machine was a foot-and-a-half-high plastic statue of the Virgin that supposedly glowed in the dark.

Not surprisingly, neither of Stephen's parents evinced similar interest in his general medical condition. My daily updates on their son's continuing deterioration were always met with the same flat, impassive stares and were always followed by the same two worried questions: namely, was Stephen still a "full code," and would I please stop all of his pain medications, "so he can more fully offer up his suffering as a sacrifice to the Blessed Virgin." Mr. and Mrs. Y.'s *only* concern about Stephen was that his pain not be ameliorated by drugs or ended prematurely by a death that a full code might postpone. In their minds, the Blessed Virgin Mary had to be appeased by Stephen's suffering. New York State law, as well as Stephen's prior wishes, guaranteed him a full code if his heart were to stop, but I steadfastly refused to withhold any of his pain medicines, a stand that his disapproving parents threatened to bring to the attention of the cardinal.

"You are persecuting us for our religious beliefs," groused an indignant Mr. Y., apparently preferring instead to let Stephen be the one who was persecuted.

Throughout the pathetic parental farce swirling around him during his final days, Stephen seemed dazed. Perhaps he was entranced by his parents' exhortations. During his brief moments of alertness, he appeared to try to understand what they were saying to him, blankly nodding assent with his father's religious admonitions. Sometimes Stephen appeared like a zombie as he stared bug-eyed into space, as his parents recited Scripture passages and prayed to the Virgin that his suffering be "acceptable" to her. Sadly, never were there any prayers for release from pain, for healing, or for love.

As these final scenes in Stephen's life droned on, I sometimes wondered what was going through his head—perhaps childhood memories, perhaps regrets about lost hopes, perhaps recollections of great opera. By that point in his hospitalization I, too, was a zombie of sorts, too emotionally spent even to care. Out of sheer self-protection, I had to feel that I, as well as the rest of the Spellman staff, had done the best we could to help him live—and die—as he wanted. Oddly, despite the freakish religious drama being played out at his bedside, and despite his increasingly unstable vital signs, I was at peace with the Stephen Y. case: it was now entirely up to Stephen. The only thing about his care I was still vigilant about was that his radio remained tuned to WQXR-FM.

Stephen Y. died on a Sunday, in the early morning hours of his 126th day of hospitalization, a week and a day after his parents appeared on the scene. His death was anticlimactic. The details were sketchy, given the late hour of his death drama. Indeed, it appeared that, in his dying, as in his living, Stephen had waited until he was essentially alone, away from me, his parents, and other familiar Spellman staff.

About twelve hours before his death, he became agitated and once again started pulling out his IVs and the breathing tube. Since his parents' arrival, he had by and large stopped such activity, but for reasons known only to Stephen, he started it up again twelve hours before the end, with an intense ferocity, like a supernova that suddenly explodes before burning out forever. Summoning up what little energy he had left, he vigorously began to fight the ICU doctor's attempts to restart the IVs and to reinsert the breathing tube. No sooner would the tube be put back in or an IV restarted than he would manage to escape his wrist restraints and yank them out again. Sedatives only temporarily calmed him down, and after a few hours of fitful sleep, he would bolt awake and resume his infernal tugging at the restraints and ripping at his IVs and other invasive tubes. He was finally able to write on his notepad "STOP" and "DNR." His parents had not visited him the previous day and could not be reached. They were at a weekend prayer vigil on the steps of Our Lady of Newark, cameras in hand, in hopes of yet another miraculous sighting of the Blessed Virgin.

Because I was out of town that weekend, the ICU doctor contacted the Spellman physician covering for me and discussed Stephen's case with her, especially noting his resistance to reintubation and his messages "STOP" and "DNR." Although these doctors could not know with certainty if he was mentally competent to opt for DNR, the commotion he was creating was clearly unprecedented, even for Stephen and his well-known previous outbursts. Indeed, something was profoundly different, singularly different, about this new situation, especially since it was the first time Stephen had insisted on being a DNR. After requestioning him about his notes "STOP" and "DNR," the ICU doctor felt Stephen understood his request for DNR. It was decided to let Stephen scribble his name on a DNR form and not to reintubate him the next time he pulled the breathing tube out. An intravenous morphine infusion was also begun, to ease any breathlessness or anxiety that might ensue. Too weak to sign his name in full, Stephen called forth his final burst of strength, printed his initials on the DNR form, pulled his breathing tube out for the last time, and died approximately forty-five minutes later.

Because no one was in the room when he died—the nurses waited until his heart monitor became flat-line before going in to see if he was finally dead—it was not clear how difficult and agonizing, if at all, his final moments were. Death for most people is usually far less terrible than the lifetime of dread and fearful expectations it engenders. The ICU staff was not able to notify his family until much later in the day; they were camped out on the church steps until late that evening, waiting for the Blessed Virgin. Stephen's body was sent down to the morgue, and his radio was finally unplugged and packed up with all his CDs, opera books, and all the religious paraphernalia brought in by his parents. The details of Stephen's funeral remained unknown, nor did anyone ever find out if his parents had witnessed that weekend of his death yet another of their miraculous visions.

A few days later, I learned from Sister Pascal that Stephen's parents had just left St. Clare's after picking up their statues, rosary beads, and photos—but not his apparently contraband opera books and CDs, which were later donated to a downtown AIDS service organization. I was relieved that they did not seek me out. I knew they would be incensed about Stephen's signing a DNR at the final moment. Sure enough, Pascal said his parents were accusing the hospital of "killing" their son by allowing him to sign a DNR, thereby, in their minds, consigning him to hell and losing their much anticipated AIDS cure by the Virgin.

Always one with just the right words of comfort, Pascal cheerfully reassured me, "Stephen's parents were looking for a miracle in the wrong place when he died. The wonderful miracle wasn't at Our Lady of Newark—it was right there in Stephen's room, where the Blessed Virgin was escorting him to heaven and to a far happier life."

Did Stephen Y. eventually come to grips with the unfinished life issues that made him cling to life so desperately, at all cost? Did he die at peace, or still with the conflicts that had plagued him for most of his life? Of course no one can say, but the healing and reparative powers of the human spirit should never be underestimated, especially at the approach of death. This self-protective, calming effort of the mind facing death has been explained in various ways. The scientist will postulate the brain's production of morphinelike chemicals called endorphins, the religious believer will evoke the presence of God and His love, and the secular humanist will try to straddle the mystical and the scientific explanations. Regardless, there appears to be something in the dying human mind that tries to put to rest any conflicts, to make sense of a lifetime of experiences. As Tolstoy wrote of Ivan Ilych, it is never too late for a person to find the "real thing" in his or her life.

What mattered about Stephen Y. was not whether he was a heroic figure or even a good person. Rather, what mattered is that he lived and died on his own terms and on his own schedule. Stephen Y. taught me to respect this most sacred of itineraries in my patients' lives, and not to rail against their wishes to linger on longer than I or other people might think necessary.

I can only hope that Stephen Y.'s death, as well as the deaths of all other Spellman patients, was as wonderful as Ivan Ilych's:

> "Death is over," he said to himself. "There is no more death." Instead of death there is light.

5

PASCAL'S PASTORAL CARE

The capacity to give one's attention to a sufferer is a very rare and difficult thing. It is almost a miracle; it is a miracle.

—*Simone Weil*

"Raphael B." . . .
"Edgardo L." . . .
"Sarah W." . . .
"Felipe A." . . .

Attending almost every memorial service is Sister Pascal Conforti, for whom death is, in her own words, an "old friend." Although momentous, Pascal's memories are indeed celebrations *of patients' courage and hope.*

EVERY WEEKDAY morning, a solitary Catholic nun of unassuming appearance leaves her convent in the South Bronx and boards the D train to Manhattan, joining the busy rush-hour crowd to Midtown. Her external attire and general demeanor are unremarkable, sometimes even a little shabby, especially on a slushy December day such as today. Indeed, her faded, blue overcoat has several small tears and unravelings on its sleeves, and her black rubber boots likewise appear old and frayed. Dressed in ordinary, decidedly unstylish street clothes, sixty-year-old Sister Pascal Conforti could easily be mistaken for a grade-school teacher, or even a nanny or cleaning lady.

Arriving at St. Clare's Hospital forty-five minutes later, Pascal first stops by the nursing office to pick up the daily census, which lists all the inpatients presently on the Spellman service, including any new admissions since the day before. Quickly dropping off her coat in her office, and allowing herself little time to warm up from the chill, she sets out on another day of pastoral care rounds, probing beyond the externality of AIDS-diseased bodies, in search of fellow sojourners in need. Indeed, the approach of Christmas has recently given special poignancy to her daily visits to Spellman's patients.

Yet despite her extraordinary ministry, Pascal is but one of a legion of largely unheralded pastoral care workers throughout the country who, like her, trek daily to their respective hospitals, nursing homes, and clinics to tend to people's spiritual needs, needs that doctors cannot, or will not, address. All too often, doctors, as well as hospital and clinic administrators, have relegated pastoral care services to the sidelines of medical care. Typically, pastoral care offices are small, nondescript cubicles squeezed, almost as an afterthought, into some obscure corner of the hospital, the more spacious quarters reserved for the new MRI scanner or the inviting outpatient medical suites so necessary these days to generate much needed income for the hospital. Pious lip service notwithstanding, pastoral care is frequently deemed a frill. Emotional support is not easily quantified on a hospital's computer printouts or quality-assurance reviews.

Likewise, many American physicians regard pastoral care as having little to do with them and their medical practices. The priest, pastor, imam, or rabbi is only someone to be called—usually by the nurses, rarely, if ever, directly by the physician—when the patient is terminal, the case "hopeless." The advent of "managed medical care," with its confusing alphabet of HMOs, PPOs, IPAs, etc., will only worsen the physician's detachment from those patient needs that cannot be met by a drug prescription or an X ray or blood test. As managed care organizations rotate patients from one doctor to another, the inevitable increase in impersonal care will risk further patient alienation and inability to face illness and even death. Physicians have never been particularly adept at dealing with death and dying, even with long-standing patients in their practices, and the cost-conscious imperative driving managed care does not encourage the doctor to spend precious time assuaging a sick patient's emotional pain or fear. In the unfolding brave new world of health care, pastoral care staff such as Pascal Conforti are going to be even more crucial in helping sick and dying patients.

In my own private practice in rural Iowa—many years before the advent of managed care—I, too, looked on pastoral care as having little relevance for me and my medical routine. My patients' spiritual needs were their own business, so I reasoned, and I was always suspicious—indeed, resentful—of attempts by the hospital priests or pastors to interject themselves into my patients' management. In an internal-medicine practice in rural Iowa, death would usually strike only the old and infirm, whose long years of life could never make death patently "unfair." Rare were the young adults succumbing to cancer or lupus or auto accidents, and their infrequent deaths were spaced far enough apart to leave my mind undisturbed by anxieties about death. Young and overly confident of medical science's claims of healing, I would probably have dismissed Sister Pascal as a pleasant but meddlesome nun who had better stay out of my way.

But on an AIDS ward such as 3A, death is not merely a frequent visitor; rather, it *hovers* there as a permanent, unnerving guest who never leaves. Thus, as an AIDS doctor, I could not so easily shrug off issues of the spirit—questions about death, about the existence of a God or the soul—as I did in Iowa. That my patients on 3A were, at first glance, "different" from me—i.e., poor and uneducated—provided only brief refuge from such questions, since my daily experiences quickly taught me that Spellman patients had exactly the same fears, hopes, and needs as I would have under similar circumstances. Moreover, I sensed that I ignored these issues of the spirit at my peril: the ingredients for burnout were definitely in place.

Helping me deal with these matters was a seemingly unlikely member of the Spellman staff, Sister Pascal Conforti, director of the Pastoral Care Service. I cannot remember the first time she was there at my side, giving me and my patients solace in her characteristically understated way. Indeed, understatement has been her hallmark, but I quickly sensed a singular strength of spirit that not only complemented my medical care but also often eclipsed it in importance. Gradually, I came to rely upon her to ease the patient pain that Percocet could not relieve, to respond to the loneliness and despair that Prozac could not reach, to understand the "difficult" patient that psychiatry could not fathom. The emotional demands of AIDS care—for both patient and caregiver—have highlighted the underlying impotence of medical care driven only by science, and pastoral care workers such as Pascal have challenged AIDS doctors such as myself to return to a more human *and* humane paradigm, to acknowledge that to treat an AIDS patient as only flesh infested by a virus is to miss the point entirely of what medicine—and AIDS—should really be about.

An unusual media event, a few months after my arrival on 3A, first showed me the unique, human dimension of AIDS that Sister Pascal uncovers daily on her rounds. A local television affiliate was visiting St. Clare's to interview a representative Spellman patient for an evening news segment on AIDS in New York City. Pascal and I were asked to chaperone the 3A patient selected, Ms. Evelyn T.

Evelyn was a forty-year-old black lady from Harlem who was recovering from what had been a severe bacterial pneumonia. Although technically a state prisoner incarcerated for drug dealing, Evelyn was on work release, a program whereby nonviolent inmates approaching parole are allowed to work in the community during the day and return to a minimum-security facility at night. Working at a garment factory by day, Evelyn had been studying evenings to get her GED. Her ultimate goal was to get a clerical job after parole, so she could regain custody of her four-year-old HIV-positive daughter, who was in a foster home. As had happened with her own mother, Evelyn had been abandoned by her husband several years earlier, and her only living family was an older sister who wanted nothing to do with her. Despite these formidable odds, Evelyn seemed resolved not to let AIDS deter her from reordering her life.

In response to the TV reporter's questions about her life with AIDS and what she would tell young people, Evelyn spoke with uncanny authority and eloquence about both her feelings and her cautionary message to others.

"AIDS," Evelyn confidently announced to the running camera, "has made me a different person. It makes me a *better* person. I don't dwell on trivial things anymore. I don't have room in my life for negative people with negative thoughts. I used to let other people put me down and tell me I'm no good. Worse yet, I used to let *me* put me down, but not anymore, not since AIDS touched me. It took AIDS to teach me I'm a worthwhile person."

The reporter went on to ask Evelyn a few general questions about her background—her family, her incarceration, her ongoing self-education for her GED—and finished by asking if she had any final comments for the TV audience. Without hesitation, Evelyn concluded with a statement that dumbfounded me as I quietly stood back in a corner of her room.

"Yes, I would just like to say that I am more than a virus, I am more than just flesh and blood. It took AIDS to teach me to respect myself, and to the young people out there I say, 'Learn my lessons without getting the virus to teach them to you!' You don't have to catch a deadly disease to find your personhood."

Evelyn's words impressed me, who, like the rest of the harried medical staff, had little time—if any—to look beyond the "flesh and blood" of my patients. Here was a woman who had never read the great philosophers or modern psychologists, but she had developed a tranquillity in life that few people, even after years of psychotherapy, attain.

After the interview, I privately remarked to Pascal my amazement—and admiration—at Evelyn's insights. Pascal, who had visited Evelyn many times in the past, concurred, but also seemed much less surprised by it all.

"Oh, I've heard Evelyn's words—in one form or another—over and over again from many of our folks here. I guess I'm just accustomed to it," Pascal replied. "It's a wonderful gift, hearing such things from our patients," she added with her smile of bemusement and grace.

This "wonderful gift" Pascal alluded to became more and more apparent as, directly and indirectly, I observed her daily pastoral care rounds. A typical day for Pascal is different from an AIDS doctor's. Unlike me, she cannot rely on a glib reply, an empty platitude, to a patient's problem. Whereas I can always at least treat symptoms, especially if the underlying disease is untreatable, Pascal risks losing a patient's trust on much more important levels. In response to a patient's complaint, I can always order a new test, or repeat an old one, but she often gets only one chance to assess a patient's problem accurately. I can, if I wish, ignore the patient obnoxiously demanding an increase in pain medicine, but Pascal dismisses a patient's emotional pain at her—and the patient's—peril. In spite of the hectic pace of my work on 3A, I gradually came to respect the pivotal role of pastoral care in AIDS care.

Indeed, beyond my own AIDS ward, the Spellman Pastoral Care Service has in more ways than one been the conscience of St. Clare's AIDS division. It is the balm, the invisible sutures, that mend the fractured spirits of both patients and staff. Although superbly assisted by Father Jack and visiting seminarians, Pascal Comforti *is* the Pastoral Care Service, having worked with Spellman's patients since its inception in 1985. During her years at St. Clare's, this one woman, it is safe to estimate, has eased the transition to death of *thousands* of AIDS patients, as well as grieved with a veritable multitude of family and friends of these patients. She has also counseled numerous staff members fighting the emotional stresses of AIDS care in the cauldronlike atmosphere of the institution. This gray-haired, Italian-American nun of spare frame and steely disposition is the true heart and soul of the Spellman Center.

Youthful in gait despite her age, Pascal dresses modestly, her only uniform a short white lab coat, white blouse, knee-length skirt, sensible walking shoes, and a silver cross necklace hanging over her chest. Her angular facial features are plain and unadorned, and her diminutive appearance is pleasantly unremarkable, much like that of a spinster Latin teacher from high school in the 1950s.

Pascal's self-effacing demeanor masks a first-class intellect. An articulate advocate of death with dignity, especially for those patients too sick to speak for themselves, she has prevailed over many an unfeeling doctor who would blindly insist on senselessly prolonging a suffering patient's life in intensive care on full life-support systems. Beneath her beatific exterior is a hard-edged New York persona that can thwart occasional attempts by patients or staff to manipulate her to their particular agendas. Moreover, she is a closet intellectual who can hold forth on weighty philosophical issues with a scholarly brilliance that astounds the Jesuit seminarians who rotate through the Pastoral Care Service. Highly regarded across the nation as one of the founders of AIDS pastoral care, Sister Pascal has written several articles on the role of spirituality in holistic care of people with AIDS, a subject of far greater complexity at St. Clare's than might first be apparent. Verily, every day Pascal's work on the Spellman service confronts her with difficult personal challenges, challenges of conscience, that epitomize the serious contradictions inherent in Catholic pastoral care in the AIDS epidemic.

Indeed, first thing this dreary December day, Pascal is faced with a patient who personifies these tensions in the Church's response to AIDS, a patient who has occupied her thoughts and prayers for several weeks. As she gets off the elevator on 3A to start her rounds, she notices a grim-faced Father Jack coming out of Raphael B.'s room, halfway down the hallway. The Christmas decorations festooning 3A's corridor starkly contrast with the almost palpable gloom around St. Clare's priest this morning. Jack has just administered the Sacrament of the Sick to Raphael and, seeing Pascal walk off the elevator, signals he wants to talk with her.

For many years in the Pastoral Care Service, Father Jack has been the diocesan priest assigned to St. Clare's. Bespectacled and a little paunchy, Jack usually eschews his clerical collar and leaves the top of his black, short-sleeved priest's shirt unbuttoned, creating a slightly frumpy, irreverent appearance. Approaching his priestly duties with the same dour, take-it-or-leave-it manner characteristic of most Catholic clerics, Jack does not suffer fools and sycophants lightly. Like many other longtime survivors of the Spellman scene, Father Jack has protected his sanity with a droll sense of humor that seems tinged with world-weary cynicism about both humankind and God. Jack's behavior contrasts with Pascal's in his willingness to express openly his opinions about the Church's response to AIDS and other issues. "All the Church seems to care about anymore are tits and dicks!" he once quipped in irritation at the hierarchy's apparent obsession with sexual matters. Always the rebel, Jack prides himself on the letters of reprimand he has accumulated over the years from bishops under whom he has worked. "One of the letters," he once boasted, "was four pages long"—and here he added with gleeful emphasis—"four pages long, *single-spaced!*"

But today, Jack has no witty remarks after his pastoral visit with Raphael B., the patient to whom he has just administered the Sacrament of the Sick.

"No change," Jack somberly reports to Pascal. "I get so fucking angry whenever I go into that room. Anyway, good luck, if you're going in to see him."

Raphael B. has recently presented Pascal with unusual challenges, challenges that severely test her religious beliefs. Raphael B., or, more correctly, *Father* Raphael B., is a Roman Catholic priest who is dying from AIDS. Thirty-nine years old and—judging from his ordination photograph—once very handsome, Father Raphael came to St. Clare's from Central America, where he had been pastor in a mountain town of five thousand people. When first diagnosed as having AIDS-related KS almost a year ago, Raphael's ecclesiastical superiors immediately transferred him from his rural parish to the small Catholic hospital in the capital city. Although at that time still fairly robust, he was strictly instructed to stay in the hospital's dormitory, to suspend any and all priestly activities, and—above all—to discuss his HIV status with no one. His quarantine was so total that he was forbidden even to go to mass; the consecrated host was always brought to his isolated room by an elderly nun, who wore a surgical mask and gloves. There was apparently no counseling, no emotional support, no expressions of caring or love: Father Raphael was an embarrassment for the Church, a problem to be dealt with discreetly.

At the time of his unexpected transfer to the capital, Raphael's parishioners were informed that their priest had been assigned to work in a TB sanitorium. A few months later, the provincial bishop solemnly told Raphael's gathered flock that their former pastor had TB, which, they were told, had—most tragically—been acquired while he was working at the TB hospital. The bishop exhorted the crowd to remember Father Raphael in its prayers.

After six months of languishing in the Central American hospital's dormitory, Raphael had received no medical care and was much sicker. He had lost forty pounds, and the progressing KS lesions had become increasingly disfiguring, with bloating of his legs that made walking painfully difficult.

Father Raphael would soon require nursing care, but none of the hospital staff was willing to touch him—*literally.* In Raphael's small Central American country, AIDS was a disease the people not only knew little about but also feared. Their archbishop had recently declared that AIDS was a punishment from God, "a mark of Cain, a warning to mankind for its iniquity."

An orphan, Raphael had no friends or family; the Church had been his only home. A sympathetic social worker at the hospital was finally able to effect his transfer to New York almost a month ago, to Terence Cardinal Cooke's AIDS Hospice, far from his homeland and far enough away to spare his superiors the final embarrassment of his approaching death.

Father Raphael's stay at Cardinal Cooke was brief: a few days after arriving, he developed shortness of breath and was admitted to St. Clare's, where tests showed he had KS in his lungs. From the outset, Raphael has remained a passive participant in his medical care, never refusing tests or treatment, but also never questioning his doctors' recommendations. Two cycles of chemotherapy have retarded the KS's progression, but Raphael's condition has deteriorated to the point he is bedridden and now requires frequent morphine injections for the increasing KS pain in his legs.

But even greater than Raphael's physical pain is his profound, and totally unarticulated, psychic pain, a desolate anguish that everyone who enters his room somehow senses from both his impenetrable silence and the stoically vacant look on his face. It is a pain no one—Pascal, Jack, the entire ward team—has been able to assuage, and the advent of Christmas has rendered his plight even more poignant for the staff, to say nothing of Raphael.

"It's so sad," Pascal responds to Father Jack this morning. "I'll go in, he'll smile and say hello, ask how I am. We might chat about the weather or other small talk, but never a word about how he feels. As soon as I ask about his home or his feelings, he says he's too tired and wants to rest. We just talk past each other, or else I just sit there and he stares at the walls." It is unusual for Pascal to elicit no response from a patient after so many visits.

Jack has likewise been unable to pierce Raphael's loneliness, even after, at Raphael's request, Jack heard his confession a few days ago.

"He needs our prayers, Sister," Jack replies, probably wishing he could say more, but scrupulously observing the seal of the confessional. "He's been abandoned by the Church he's devoted his life to." Jack's voice starts to rise in anger, but then calms down somewhat. "Anyway, he's been so beaten down, so made to feel like a piece of shit, there's probably not much we can do to undo the damage. Even the sacraments don't seem to ease the pain, the guilt. Anyway, I hope God takes him soon, so his soul can finally be at peace." Shaking her head in knowing commiseration, Pascal sympathetically pats Jack's shoulder and heads into Raphael's room.

Father Raphael, like many times before, immediately begs off from any conversation, claiming to be too tired to talk.

"But please stay for a while, if you'd like," adds Raphael, always polite to his few visitors. Soft-spoken and gentle in manner, Raphael always seems embarrassed by his very existence—*a priest with AIDS,* who cannot claim to have contracted the virus from a tainted blood transfusion. "I just don't feel like talking today, Sister. I hope you understand."

"No problem, Father," assures Pascal, trying hard to act blasé. "I'll just sit here and rest my weary feet for a few minutes."

Father Raphael actually does look tired today, even a little breathless, for the second course of chemotherapy has not been as effective as the first. Bedsheets cover his swollen legs, which have ballooned to at least twice their normal size. For several days, clear lymph fluid has been oozing through the breaks in his fragile, fissured skin, staining the overlying sheets. Multiple, large purple blemishes stud Raphael's gaunt face, and a large lesion on the tip of his nose is starting to ulcerate—undoubtedly the "mark of Cain" in his archbishop's eyes. The chemotherapy has predictably thinned his head of thick, jet-black hair, and his bony shoulders now support only vestiges of once impressive muscles, acquired from years of helping his parishioners, all poor peasants, with their farmwork during busy harvest seasons.

The only personal item in Raphael's room is a curious one, which he brought with him from Cardinal Cooke Hospice. On the bedside stand, turned toward him, is a framed photograph of him in his midtwenties, in clerical garb, next to another young priest. Both men appear happy and robust, and Father Raphael has his hand on his friend's shoulder. Pascal once inquired about this photograph, only to be told it was taken at his ordination and that "it was all a long time ago . . . I really don't want to talk about it." Indeed, as he waits to die, Father Raphael does not want to talk about anything, preferring to gaze quietly out into his room with a forlorn sadness that denotes neither fear nor anger nor self-pity. Rather, his blank, melancholic stare seems to epitomize unspoken regrets.

Fragmentary records accompanying Raphael to America indicate he had been raised in a Church orphanage in a provincial town, and presumably because of precocious intelligence and innate spirituality, he had been sent to the country's only seminary, in the capital city, where he was ordained eight years ago. After ordination, he briefly studied in Rome, where he learned English and was offered the opportunity to pursue an advanced degree in theology at the Gregorian University there. However, Raphael apparently opted to return to his native country, to work among his own people.

Although these records did not elaborate, and Father Raphael never broke his self-imposed silence about his past, one of Father Jack's clerical friends, who has worked in Central America, was quite familiar with Father Raphael's country and had given Jack and Pascal tantalizing background information. Raphael's parish apparently included several small villages that surrounded the largest town of five thousand, which had the parish church. As priest, Raphael was the religious and moral authority for the area, not only performing the usual Catholic sacraments—"hatch, match, dispatch," as clerics waggishly encapsulate birth, marriage, and death—but also mediating family conflicts, being a village advocate with the local military government, and comforting the sick and dying.

"Father Raphael didn't have an easy job, I can tell you that," this priest friend of Jack's concluded. "Oftentimes these village priests are *it*—psychotherapist, social worker, doctor—you name it. The Church's been in a lot of turmoil over the years there. Their archbishop's even too reactionary for Rome—if you can believe it—and the countryside's been festering with liberation theology. Sometimes the rural clergy is at the forefront of reform; other times, the village priests are so isolated they become more conservative than the hierarchy. It'd be interesting to know which side Father Raphael was on."

Father Raphael's vow of silence was impervious at St. Clare's.

Today, as Pascal sits at the bedside of this AIDS exile, the sepulchral silence in Raphael's room compels her to confront painful issues she has often had to face over her years at St. Clare's. Many believe the AIDS crisis has been the greatest challenge to the Church's moral authority since the Reformation. Raphael's fragmentary story has troubled Pascal, raising haunting questions she senses will never be answered. What does Raphael feel now about his faith, about God and the Church? How could his country's Church treat him the way it did, and what does this treatment say about the healing mission she feels the Church must have toward AIDS?

These musings at Raphael's silent bedside remind Pascal of the uniqueness of Catholic pastoral care in the AIDS epidemic, especially at St. Clare's, where tensions have always existed between medical realities and religious beliefs.

Indeed, Catholicism's contradictory response to AIDS is nowhere better illustrated than at St. Clare's Spellman Center. The very name of the hospital's AIDS center is both supremely ironic and cryptically penitential, perhaps an attempt at contrition for past sins of the Catholic Church in New York. Indeed, Francis Cardinal Spellman, archbishop of New York from 1939 to 1967, had been a close intimate of rabidly homophobic, deeply closeted politicos Roy Cohen and J. Edgar Hoover. The credibility of the New York hierarchy was further strained in the 1980s when many of its younger priests were dying of a mysterious illness that Church leaders described as "liver cancer" or other more palatable maladies. To admit that its priestly ranks were being decimated by the new plague called AIDS, the Church would have had to acknowledge the unmentionable: that sizable numbers of gay priests acted on their sexual orientation. Unlike in Father Raphael's country, the American Church could not indefinitely ignore the AIDS epidemic, and its response has spawned contradictions between everyday practice and doctrinal principles—contradictions that flourish at St. Clare's.

To an outside observer, St. Clare's appears typically Roman Catholic, with all the trappings common to countless Catholic hospitals large and small across the world. Indeed, the symbols of Catholicism are so universal that Father Raphael has undoubtedly felt right at home at St. Clare's. There is a jewel-box chapel, where noon mass is celebrated daily by Father Jack. Public hallways and patient rooms are adorned by a host of somber crucifixes and pictures of the Blessed Virgin Mary. At least a half dozen marble statues of the Virgin, St. Luke, patron saint of physicians, and St. Clare fill pedestals and niches in the hospital's public areas. Several of these statues are colorfully decorated with candles and votive offerings. And emphasizing a more contemporary association with the Church, in the main lobby is a large, framed photograph of His Eminence, John Cardinal O'Connor, vested in the crimson regalia of his sacred office and shown beneficently smiling down upon his flock, with a frozen gesture of benediction—an archdiocesan icon, a sacred epiphany of Catholic munificence toward those suffering with AIDS.

John Cardinal O'Connor himself has embodied the contradictions in the Church's response to AIDS. This good and holy man is neither the faultless saint his apologists claim nor the devil incarnate his rabid critics like to label him. Had O'Connor been Father Raphael's archbishop a year ago, Pascal's bedside visit today would be much different, efficacious and healing. As archbishop of New York, Cardinal O'Connor is the ultimate boss, the CEO of

St. Clare's. Once a year, His Eminence is escorted on a choreographed tour of the hospital, where he dispenses his blessings on admiring staff, quizzical patients, and obsequious administrators. A tireless spokesman for strict orthodoxy as promulgated from Rome—he is *highly* regarded by His Holiness the pope—Cardinal O'Connor has been an especially controversial cleric in New York, where the archbishop has always been a powerful political, as well as religious, figure. His unbending views on homosexuality, abortion, and birth control, including condoms, have earned him both enmity and praise from large segments of the city. However, O'Connor has also been unstinting in his support of the healing ministry of the Church in treating people with AIDS. He may be opposed to gay rights and sexual practices, but he is definitely not AIDS-phobic. When His Eminence once wrote to a Spellman doctor the tribute "You humble me," he was sincere in his humility.

The cardinal would probably just as soon prefer not to deal with the vexatious problems of managing a largely unprofitable archdiocesan hospital such as St. Clare's. Moreover, although never admitting it, O'Connor would also prefer not knowing about the *major* deviations—indeed, flagrant violations—from Church teachings that occur daily at St. Clare's Spellman Center. For, not only is safe sex candidly reviewed with patients in the Spellman outpatient clinic, but the staff there also commits even graver transgressions of Church doctrine behind the closed doors of the clinic's exam rooms, safely away from the glare of Church censure. Sometimes the challenges to Church teachings are even more public: several times a year, Spellman's weekly medical-education conferences feature sociologists and psychologists who discuss with the staff the psychosocial dynamics of safe-sex education and condom usage. One recent conference even hosted a city-sponsored acting troupe whose members demonstrated the explicit safe-sex skits they had been performing for inner-city schools, complete with coarse street talk that would not amuse His Eminence. Finally, the majority of the Spellman staff, many of whom are gay, have much more liberal views than the cardinal and have been openly critical of his doctrinal teachings. It is indeed unfortunate that Raphael B. is too far gone, both physically and emotionally, to appreciate the unabashed unorthodoxy churning around him daily in this Catholic AIDS service. Such a realization might perhaps have encouraged him to voice his inner feelings, unorthodox as they might be.

As she continues to sit at Father Raphael's bedside, Pascal is probably pondering personal questions about this priest with AIDS. Namely, how did

Raphael acquire HIV, who is the other young priest in Raphael's ordination photograph, and why was it the only thing he brought with him to St. Clare's? And, most tantalizing, what was the "guilt" Father Jack, perhaps a bit too loosely, spoke of a few moments earlier in the hallway? Although her visits over the past weeks with Father Raphael have largely passed in silence, Pascal has nonetheless sensed in his few words to her a gentleness, even a playfulness, of spirit that might have made him a valuable pastoral care colleague at St. Clare's, especially in the quirky milieu of Spellman's doctrinal pluralism. Such fanciful conjecture should lighten Pascal's otherwise serious musings today—Father Raphael as a Spellman priest might have *really* given Father Jack some competition in stepping on the cardinal's red dress, given the wackiness of the Spellman service at times.

The cardinal, of course, is no fool. He knows about the heretical goings-on at St. Clare's, as elsewhere in his diocese, but unless his nose is rubbed in it, he is too busy with more pressing matters of Church policy. Every so often, he has to respond—almost perfunctorily, so it seems—when a particularly lurid departure from orthodoxy at Spellman is reported to him, such as referring a patient to Planned Parenthood. The Spellman medical director will then quickly smooth things over and urge the staff to be a tad more discreet in ignoring Church policy. Within a brief time, things then return to the way they had been.

Cardinal O'Connor's ambivalence, perhaps even tolerance, toward the Spellman Center's doctrinal waywardness reflects not only pragmatism but also acknowledgment of an essential Catholic precept that guides frontline workers such as Pascal and Father Jack, especially when their faith is challenged by cases such as Father Raphael's. Highlighted by Pope John XXIII's Vatican II Council, this guiding precept is really as old as Catholicism itself—namely, all Catholics are *obliged* to follow their consciences in all things. Thus, with regard to Church teachings on sexual issues relating to AIDS, a Catholic can prayerfully dissent from such teachings and can even, within limits, act on this dissent. Where the cardinal takes issue with this kind of dissent is when it leads to guest spots on *Oprah,* op-ed pieces in the *New York Times,* or ACT-UP demonstrations in the cathedral. Indeed, despite his public persona as a prince of the Church, the cardinal's private reactions to digressions from Church teachings about safe sex and condoms might be surprising, since he deeply understands the moral dilemmas such teachings impose in everyday practice. One wonders, for example, what his response would be if, in the absolute privacy of the confessional, he were confronted with an African missionary—or a St. Clare's staff member—who admitted to privately sanctioning safe sex and condom distribution to prevent the spread of AIDS. If he were in Pascal's place today, at Father Raphael's bedside, O'Connor's personal feelings about Raphael B.'s situation would probably surprise both his vocal critics and his ardent supporters.

Pascal herself has never openly criticized the Church hierarchy about anything, including its approach to AIDS and other social issues, although one senses she holds strong opinions on many matters. Always the diplomat, she will deftly change the subject whenever anyone tries to engage her in debate of such controversies.

"I'm busy enough with my work at Spellman," she begs off. I believe that she is silently confident that, if Jesus were alive today, he would be at St. Clare's, not at St. Patrick's.

The unresolved tensions and schizoid contradictions inherent in St. Clare's Spellman Center are neither hypocritical nor laughable. For Catholics such as Pascal and Jack, they are serious matters that always demand prayerful reflection, as Pascal is doing today in Father Raphael's room. Undoubtedly, Pascal fears that Raphael's apparent estrangement from God and humankind stems from the guilt that often results from the conflicting interplay of faith, dogma, sex, and sexuality. Father Raphael's desolation seems to bespeak a soul that believes it is beyond redemption and unworthy of peace.

Pascal's musings about Father Raphael and Mother Church make her shudder. After saying good-bye to him—he smiles back but is strong enough only to mutely mouth his farewell—she leaves to continue her rounds.

Daily census in hand, Pascal scans the list for any new Spellman admissions from the night before. Today, the only new patient is Georgie J., who, by virtue of previous admissions, is well-known—some would say notorious—to both Pascal and the ward staff. Although in the hospital barely twelve hours this time, Georgie is having one of her infamous hissy fits, working both herself and the staff into hysterics. Indeed, Georgie's case graphically illustrates how pastoral care can diffuse misunderstandings between patient and staff, misunderstandings that threaten to interfere with the patient's medical care.

A twenty-eight-year-old partial transsexual of mixed Korean and Argentinian ancestry, Georgie has come in with yet another bacterial pneumonia, resulting most likely from shooting coke and heroin into her neck veins, the only ones still usable after many years of drug abuse. A self-described "chick with a dick," Georgie supports her drug habit by hustling the Times Square area, bedecked in a rusty red wig, supertight miniskirt, and perilously high heels. Several times she has been mauled by johns enraged to learn they had been tricking with a guy with big breasts instead of a girl. Whenever she was hospitalized at St. Clare's, Georgie would have frequent outbursts at the slightest provocations, often refusing medical care and lashing out at the ward staff for no apparent reason, later to apologize contritely to the recipients of her tantrums. Most of the Spellman staff realize that Georgie really means no harm, despite her volatility. One moment she can be your worst enemy, blowing you away with bitchy invective, and the next moment, your best friend. At her calmest, Georgie is histrionic and theatrical, and when overly excited—which is her usual baseline—she can be totally out of control, dramatically flailing about in extravagant hysteria that has become her trademark.

Today, Georgie is at peak form as Pascal heads toward her room to say hello to her "old friend." She and Georgie go way back from many prior hospitalizations, when, during periodic raptures of short-lived piety, Georgie was, at her own insistence, "baptized" *four* times in St. Clare's chapel. In fact, Father Jack had baptized her only once and at subsequent ceremonies was able to preserve Catholic dogma by solemnly anointing her with holy oil in the Sacrament of the Sick, which has always provided her sufficient reassurance of salvation.

But today, "Blessed are the meek" and "Turn the other cheek" are the dictums furthest from Georgie's mind. She is angry, royally annoyed, with me, her doctor. When I examined her an hour earlier this morning, I had the temerity to raise with her the subject of CPR; that is, what, if anything, would she want to have done if her pneumonia, which extensively involves both of her lungs, worsens and threatens to kill her. I can sometimes be too matter-of-fact—some would even say blunt—in my assessments of how sick a patient is, and such discussions of CPR issues have occasionally blown up in my face, as they did earlier today with Georgie.

When I thus reviewed CPR/DNR options with her, Georgie made the mistake of anxiously asking me, her voice dripping with histrionic angst, "I'm not going to *die,* am I?"

Impatient with her theatrics—I have tangled many times with Georgie on past hospitalizations—I unwisely shot back, *"Yes,* Georgie, *you're* going to die, *I'm* going to die, we're *all* going to die. I don't *think* you're going to die from your pneumonia, but there's no way to be sure about anything. I could get hit by a bus tonight, your pneumonia could get worse, or a terrorist H-bomb could blow up New York at high noon today. You have to face the fact you're sick, or else you'll *really* freak out when you eventually do face death."

Apparently such realities were too much for Georgie to assimilate so early in the morning. Right after I finished this scolding, her eyes became as big as her gold-plated bangle earrings, her mouth dropped wide open, and her face was transfixed—frozen—with sheer horror at my declaration of her mortality. As she had frequently done on our past confrontations, Georgie suddenly appeared catatonic, unable—or unwilling—to talk further with me. After quickly making sure she still had a pulse—she did, and it felt seething—I warily retreated from her room, certain she would make me pay dearly for this faux pas, perhaps by refusing to take her medicines or to go for a necessary test. Georgie and I had often had our moments over the years, our relationship alternating between love fests and heated arguments. Indeed, as soon as Pascal arrives on the floor this morning, I ask her help in once again smoothing things over between Georgie and me.

"Georgie's on the warpath again, Sister. Her pneumonia's pretty bad this time, and all I wanted was just to get her to *think* about DNR, and she goes and freaks out on me. It's the silent treatment again." Pascal's willingness to intercede is, as always, taken for granted, and as she enters Georgie's room, she is confident she can quell this mini-tempest, given their history of mutual friendship and civility.

"Sister!" Georgie urgently cries out in a thick Latin accent as soon as Pascal steps into her room. "You know what Dr. Baxter say to me?" Before Pascal can even breathe in to reply, Georgie excitedly clamors on, flailing her arms about for dramatic effect.

"He say I going to die, he say my pneumonia get worse, he say it kill me, he say *everyone* going to die! *Tell me it not true, Sister!* Tell Georgie she not die!" Her tone turns more irritated. "I *very* angry with Dr. Baxter. Why he say that to me, Sister? Georgie no like Dr. Baxter no more. Dr. Baxter *mean!"* Georgie scowls as Pascal listens with an air of equanimity and gently nods in understanding but not assent. She knows this outburst is more a minor squall than a full-blown gale.

"Wasn't Dr. Baxter your doctor the last time you were here, when you had to go to intensive care for your pneumonia?" Pascal's inquiry sounds innocent enough, but she is really trying to remind Georgie that Dr. Baxter is not a stranger but really an "old friend," as well.

"Sí, but why he say that to me? Why he so mean to me? Georgie no more talk to Dr. Baxter!" Already Pascal's presence is taking the edge off of Georgie's anxiety attack.

"Maybe Dr. Baxter didn't mean to upset you so. Maybe he was worried about you, like he was the last time you were sick. You know how Dr. Baxter likes to worry." To the latter remark Pascal adds a little shrug and smile, as if again to suggest that everyone on Spellman is really family, all working to the same end.

"But, Sister, I so a'scared, and Dr. Baxter say I going to die. He scare me with all his talk of death . . . death, death, always death! All I hear from Dr. Baxter is *death, death, death!"* Georgie sits up in bed and solemnly flings her upper body about in campy theatrics, like an overripe meld of Norma Desmond and Carmen Miranda.

Pascal nods again in sympathy but senses that even Georgie knows her anxiety is overblown. After a brief silence, Pascal asks, as if nothing has happened, "So how have you been recently?"

Georgie blithely skips over Pascal's query and, as on past confrontations, suddenly shifts to a tone of contrition and self-confession. "You know, Sister, I think about what Dr. Baxter say." Her face lights up with an impish smile, her eyes affectionately look up at Pascal. "Dr. Baxter speak the truth to Georgie about death. *Sí,* Sister, he speak the truth. He scare Georgie, *sí,* but he also speak the truth."

Pascal's nod conveys agreement, not just empathy. "Tell me, Georgie, what do you think about what Dr. Baxter said?"

For the next ten minutes, Georgie tells Pascal of her life story, her flamboyant gestures somehow leavening, as if in counterpoint, her trail of pain. Always attentive, Pascal has nonetheless heard Georgie's tale several times before—the story of her sexual abuse by her alcoholic stepfather, her frequent trips to jail for prostitution and other petty crimes, her decision to have breast implants "so Georgie look like *real* woman," her innumerable abusive boyfriends, her flights with drugs, her dangerous but proud life as a street hustler—"Your Georgie the *prettiest* hooker in Times Square, Sister"—and finally, her fears about how AIDS is becoming more and more difficult for her to ignore and may someday even prevent her from working for a living.

"I know someday Georgie die." Fear partially comes through the theatrics. "I want you be there when I die, Sister. You be there with your Georgie when she die, Sister?" Georgie, a thespian to the core, emphasizes the urgency of her request by dramatically reaching out for Pascal's hand. But before Pascal can reply, Georgie quickly switches to a recurrent refuge from her worries.

"I want baptize again, Sister," Georgie declares confidently. "Tell Father Jack, 'Georgie want baptize again.' God forgive His Georgie if she baptize again." Yet her assertion that she will be forgiven has a tinge of uncertainty. Pleadingly, she looks to Pascal for validation of her contention "God forgive His Georgie."

But Pascal needs no cue. "Georgie, God has already forgiven *all* of us"—here she adds a playful wink—"and I happen to know from the highest sources He has a special soft spot in His heart for Georgie J."

Growing tired from conversation—she never did tolerate much introspection—Georgie is now teary-eyed, her contentious facade having dissipated in Pascal's presence. Pascal holds Georgie's hands, which are scarred from ancient needle tracks.

"Georgie, you really are a *wonderful* gal, and we can't afford to lose you. This place wouldn't be the same without you, and we're not about to deprive Times Square of its prettiest girl, with her big heart!" Pascal is genuine. "We won't lose you!"

As Pascal waves good-bye, Georgie yells after her, "Tell Dr. Baxter I no hate him. Tell him he speak the truth."

Next is Israel T., who has been in the hospital for almost four weeks, and who illustrates the insoluble emotional problems pastoral care workers such as Pascal must confront daily.

A forty-one-year-old Hispanic released from prison three months ago, Israel has been recovering from a bad episode of PCP, which was so severe that it looked as if he might not make it only a week ago, when all his attentions were centered on just getting enough air from one breath to the next. But now that his acute medical problems are receding, Israel's concerns have once again returned to his daunting personal problems.

"I'd rather fight the PCP than have to face my life right now," he told Pascal on her rounds a few days ago. Because of his protracted illness, Israel is again homeless. The Midtown SRO hotel room assigned him after release from prison was broken into, and his few possessions were stolen during his first week of hospitalization. Then, as his stay at St. Clare's dragged on, welfare regulations caused him to lose his place altogether. Furthermore, he is once again feeling his old hunger for crack; he briefly went AWOL for drugs earlier during this hospital stay, but incapacitating breathlessness quickly brought him back in.

But these are familiar—indeed, routine—problems for Israel, who is now tormented by a newer, even greater worry. After several days of frantic calls from the unit's only pay phone, he has finally learned that his six-year-old son, Samuel, whom he has not seen for five years, is suffering from cystic fibrosis and is in a hospital in Buffalo. The doctors there will not give out any information because his ex-wife, Maribel, has forbidden it. For the past several days, Israel has been obsessed with concern about his only child, as if his own survival were linked inextricably to his son's condition.

Today Pascal has both good and bad news, but she has decided to tell him only the good:

"One of the sisters in my convent knows a social worker in Samuel's hospital. She'll try and see if Maribel can be convinced to give out information on Samuel. She'll let us know as soon as she has any news." Pascal does not tell Israel that the word from Buffalo is bleak: Samuel is critically ill, tethered to a breathing machine in intensive care, brain-dead and not expected to live beyond the week. Not only is this news still confidential, per the mother's wishes, but also there is nothing Israel can do for now anyway—he is still too sick to travel to Buffalo, even if he could afford to, which, of course, he cannot.

"Sister, I gotta see my boy before I die," Israel forlornly explains today, oblivious to the fact that, sick as he is from AIDS, he will probably outlive his progeny. "I don't even know if he knows I'm alive . . . my own flesh and blood, and he don't even know I'm alive." A sadness mingles with bitterness in his voice—nothing is worse than the death of one's child.

"I never meant for things to turn out this way," he continues, more to himself than to Pascal. "I know I got a drug problem, but I ain't no junkie, Sister. I swear to God I ain't never shot up for fifteen years. No needles for fifteen years," he mutters to himself, "for *fifteen years* and I still get the fucking virus! It's not fair!"

Pascal has heard Israel's lament many times before, but she patiently listens with anticipation, as if she were hearing it for the first time.

"It's just not fair," Israel goes on. "I get busted upstate for five years and the guy I sold the stuff to gets a fucking suspended sentence 'cause he's a big-shot doctor—*a doctor* for Christ's sake!" Bitterness has yielded to anger. "And I can't even see my own son. *It's not fair.*"

Six feet two inches tall, with a generous mustache and long brown hair, Israel, despite his illness, remains a handsome, muscular man, not at all emaciated like many of his fellow patients. Indeed, because of this robustness, he has often gone AWOL during prior hospitalizations, without completing treatment. Israel has always been suspicious of his caregivers at St. Clare's. He rarely acquiesces to lab tests or X rays the first time around and has to be asked a second or third time before he reluctantly allows them. Although never overtly hostile to his doctors and PAs, he always projects a distrusting air, rarely saying much of anything on medical rounds. Seldom does he volunteer any symptoms, even when he is sick, and personal feelings never cross his lips when doctors visit him.

But Pascal's rounds are occasion for Israel to voice his feelings, perhaps because she never wants anything from him—not permission for a medical test, not agreement to take medication, not even willingness to pray or even to acknowledge a God. Pascal is just there, resisting any misguided urges to speak the right words of comforting maxims.

The mere fact that Israel is confiding in Pascal can be regarded as a minor triumph. He has lived a lifetime of being told that his feelings were bad or, even worse, unimportant. This nun simply listens and accepts, unconditionally. Perhaps Israel responds to Pascal's total lack of pity. As she would remind the more piously inclined, "Pity distances people. It misdirects our love and separates us from those we seek to serve."

Today, as on prior visits, Israel cannot move to a higher plane—he cannot rise above the bitterness, the feelings of injustice. Sitting on the edge of his bed, he stares down at the floor, brows furrowed in anger, muttering over and over to himself, "It's not fair, it's not fair . . ."

Concluding her visit, Pascal stands beside Israel and softly touches his shoulder and, eyes partially closed, mouths to herself the briefest of prayers, literally only a second or two in duration. Israel does not hear her words but knows what she is doing. The most minuscule of tears begin to well in his eyes, and he nervously changes the subject.

"Could you please get me a get-well card so I can send it to Samuel," he asks, handing her a dollar. "Something for a kid who's six."

"Israel, consider it done. And let me tell you something: don't ever let anyone tell you you're not a loving father. You're a *great guy,* and Samuel is *darn lucky* to have a dad like you."

Israel does not reply. A look of profound desolation has replaced his anger.

As she is exiting Israel's room, Pascal almost literally runs into Lois M., a patient who can be regarded as a pastoral care success story.

"Good morning, Sister," Lois cheerfully greets her friend. "I'm more sure than ever about what we talked about yesterday." Speaking in hushed tones, Lois seems shy about letting anyone overhear the big news she confided to Pascal the previous day.

Lois's animated, friendly manner belies how sick she was when she was admitted to Spellman two days ago for severe frostbite of her feet, acquired from falling asleep overnight in a cardboard box on a heating grate behind the Plaza Hotel. At that time, her feet were red, swollen, and tender to the touch—bearing the slightest weight caused her exquisite pain. But, thankfully, today she can walk without the limp of only a few days ago, and it appears that she will eventually be able to leave the hospital with both feet, including ten toes, intact.

This St. Clare's admission was Lois's first. Like many Spellman patients, Lois has far more urgent problems in her life than her AIDS, which has largely been asymptomatic over the three years since she tested positive. Yesterday Pascal visited her for the first time, and unlike many new patients, Lois was more than eager to recount her life story: her abandonment by her mother at age seven, the subsequent sexual abuse by her stepfather and two brothers, her being shunted from one foster-care home to another, her chronic alcoholism, her prostitution to support herself, her six months in jail at Riker's Island, her recent eviction from her "home" in the women's bathroom of the Amtrak train station, and—her big secret—her anxiety in learning last month she is pregnant for the first time, a pregnancy that was unplanned. The prospect of having a baby has given Lois a completely new perspective on her life, even motivating her to maintain a shaky sobriety for the past several weeks.

"I always wanted a baby, someday," Lois tearfully explained to Pascal yesterday, "but not right now, not the way my life is now." Pascal listened intently but said nothing in reply.

"I want to get my life together first," Lois continued. "Now I have to deal with AIDS and booze *and* a baby? I don't know if I can do it. I don't want to kill the baby, and I don't want to give it up. I always wanted a baby, but I don't know if I can do it, Sister Pascal . . . I just don't know. I *really* do want to keep it, but it's not the right time, Sister."

Pascal's conversation that first pastoral visit reflected the fact that Lois was deeply conflicted in her hopes and fears. Yet despite her misgivings and her own health problems, Lois conveyed yesterday a primal imperative of motherhood. "I'm pretty sure I want my baby; no, I'm more than 'pretty sure.' I *do* want my baby," she firmly concluded. Her "conversation" with Pascal had largely been a monologue whereby her true feelings emerged.

"Yes, Sister, I *do* want to keep my baby," she relayed to Pascal yesterday, who finally broke her silence and responded in kind, *"So you shall,* and it's going to be a *beautiful* baby!"

Nineteen years old, seriously malnourished, and registering a T-cell count of 2, Lois M. probably does not have much time left, but for reasons only a homeless HIV-infected young woman can truly understand, she is determined to have her baby, despite the odds against both of them. Lois might not even live to see her baby's first birthday, and despite recent medical advances, there remains a real risk the baby itself might be born HIV-infected. But Lois's decision reflects issues of mortality and immortality—above all, issues of hope—that can be appreciated only through the special prism of AIDS, which often crystallizes life's meaning for people like Lois in a short time.

This morning, the change from the previous day is pronounced. There is no longer any tearful introspection. Appearing totally composed, Lois has made up her mind.

"I've talked with the doctor, just like you said," she excitedly whispers to Pascal in the hallway, "and he's going to refer me to the OB doctor in the clinic . . . and Abby A. is trying to get me into Gift of God, so I can have a place to stay after the baby's born." It is difficult to tell whose face is more transfixed with a look of unmistakable happiness—Pascal's or Lois's.

One can only surmise that Lois's search for a meaningful existence, even a feeling of immortality, is now linked unequivocally with the fetus she is carrying. For the first time in a long time, Lois M. feels her life means much more than alcohol and AIDS.

After promising to return tomorrow to talk more, Pascal gives Lois a hug and heads down the hallway to see Edgardo L., another "old friend," as she always calls the patients she has followed over the months and years of their HIV disease. Like Georgie J., Edgardo's case reflects how Spellman's pastoral ministry has accepted lifestyles otherwise condemned by the Church.

As she quietly enters his darkened room, Edgardo is barely conscious, staring blankly at the cracked ceiling overhead. He is slowly slipping away from complications of a bacterial pneumonia. His vacant, glassy-eyed gaze evinces a brief, almost imperceptible glimmer of recognition when Pascal approaches his bedside and greets him. He moves his gaping, parched mouth ever so slightly in acknowledgment of her hello. However, this spark of sentience quickly fades back to a premorbid mask of frozen terror.

Edgardo L. has been a celebrity patient, at least by Spellman standards. A sixty-eight-year-old Mexican, Edgardo was a Marxist poet of minor renown in the late 1950s, and during his early years as an intellectual revolutionary, he had personally known Fidel Castro, Che Guevara, and even Hemingway, having met him in Havana. Edgardo's celebrity was brief, and up until his illness, he worked as a waiter at a Midtown coffee shop owned by his male lover of many years. Never fully accepting his sexual orientation—he was a typically "tortured" Latin-American homosexual—Edgardo would always assure Sister Pascal, as if it mattered to her, that he had always been "chaste" with his lover, from whom he has recently been estranged. Floridly theatrical in his pronouncements, Edgardo always spoke with a thick Latino accent that often made it difficult to understand his English, despite decades of his living in New York.

With aquiline nose and graying, wavy hair, Edgardo had always maintained an aloof air of genteel, albeit impoverished, aristocracy, even in the face of advancing emaciation from AIDS. His proud, mercurial personality often alienated even his closest friends, all of whom were ultimately driven away as he became sicker and more demanding. A neighbor across the hallway helped him at his studio apartment toward the end of his HIV illness. A retired black sanitation worker, this neighbor was not really a personal friend, but he nonetheless assisted Edgardo whenever he could, especially with shopping and looking after his apartment during his hospitalizations. The succinct reason this neighbor once gave for his generosity was one that only a "real New Yorker" could fathom: "He was alone and needed help." This largely anonymous Good Samaritan recently described Edgardo to Pascal as "a tough case—one moment he's your best friend, the next moment your sworn enemy." Perhaps the dissimilarity, and lack of personal bond, between Edgardo and his neighbor allowed them to coexist. And once Edgardo was hospitalized, this neighbor visited him only a few times, relinquishing his care more out of respect for Edgardo's privacy than lack of concern.

"Nothing is worse than a dying artiste," Edgardo once histrionically reminded Pascal, evincing a self-awareness that impressed his caregivers. Indeed, his emotional response to the vicissitudes of AIDS has amply demonstrated this axiom. Alternating between terrified whimpering and imperious rage, Edgardo has given up on living during this hospitalization and has been content to stew away in bed, never trying to help himself. In fact, he has refused to "fight" the pneumonia, which did not necessarily have to be fatal this time around, given the likelihood it would respond to antibiotic therapy. Instead, for reasons unknown to everyone, Edgardo has taken to bed in a spirit of defeat and hopelessness. Neither psychiatric intervention nor Pascal's best efforts have been able to reach him.

As he gradually weakened from the pneumonia, Edgardo has steadfastly refused to discuss his feelings about his life and his illness, preferring instead to speak in cryptic generalities, often highlighted by dramatic gestures and flourishes meant to camouflage rage and self-loathing. Although repeatedly told he is "giving up too soon," Edgardo refuses to internalize such advice. Like Lois M., he has made an unalterable decision.

Pascal's visit today is partially prompted by an alarming report she received from the unit's nurses. Apparently, Edgardo's ex-lover visited him late yesterday afternoon and created a major commotion in his room. According to the floor staff, the ex-lover could be heard, through the closed door of Edgardo's room, contemptuously berating him about his funeral.

"That's just great!" the ex-lover screamed as Edgardo lay helpless from his pneumonia. "You told me you had ten thousand dollars in the bank, but now you only have six thousand! It's going to cost me four thousand dollars to bury you! How could you be such a fucking shit! How can you just lie there and say nothing! You can forget about your bones *ever* returning to Mexico to be with your mother's!" The ex-lover then loudly slammed the door as he stormed out of the room and off the floor. Although lethargic, Edgardo most likely comprehended the words.

Today, as she sits next to Edgardo's bed, Pascal holds his bony hands and asks how he is doing. Undeterred by his apparent lack of response, she continues on, in her matter-of-fact way, as if she were just making a routine visit with an old friend.

"You know, Edgardo, last night I read a translation—a very good translation, mind you—of your first book of poems, and, let me tell you, what a wonderful gift you have." Edgardo still stares blankly at the ceiling, his breathing slightly labored and stertorous from the desiccated phlegm accumulating in his throat. Edgardo wrote his first book of poems in protest of Cuba's Batista regime, four decades ago.

"The poem about the caged bird that's accidently freed—I think it's entitled 'The Soaring Spirit of the Dove'—anyway, I loved your metaphor of the caged bird and how it represents the imprisoned soul of mankind." Then, slipping in an interpretation that Edgardo probably did not have in mind when he wrote his poem many years ago, Pascal adds, with seamless continuity, "It's wonderful, isn't it, to think of our souls as beautiful birds released to the freedom of a far better world God has for us."

Edgardo does not acknowledge her words, but it is likely he heard them, just as he comprehended the wrath of his former life partner yesterday. Pascal quietly sits with this wisp of a man, who remains alone with his thoughts, emotions he has been too proud—or too ashamed—to share with anyone. He is too weak even to clear the phlegm from his throat, let alone talk. The time for giving voice to these thoughts has passed, and her presence is the only remaining comfort Pascal can give him.

Pascal barely closes the door on her way out of Edgardo's room before she walks straight into Rita B.'s big bear-hug embrace.

"God bless you again, Sister," Rita B. booms, delighted to have run into Pascal. "I don't know what I would have done without you!" Her husky voice reverberates throughout the hallway.

Rita is on her way home, having just been discharged after a week's stay on Spellman. A hefty, forty-five-year-old black woman, Rita was HIV-infected some time ago by a passing boyfriend, whose memory has faded into oblivion. This present admission, her first to Spellman, was for acute pyelonephritis, a severe kidney infection, which promptly responded to intravenous fluids and antibiotics. Like many other Spellman patients, however, Rita's major concern—her all-consuming worry during this admission—was not her HIV disease, but her eighty-one-year-old infirm mother, whom she had to leave alone in their small apartment in a high-rise, high-crime public-housing project on the Lower East Side. Without friends or family to help out, Rita and her mother have only themselves.

A week ago, Rita's voice was far more muted than it is today. After an ambulance had rushed Rita to the hospital with high fever, severe flank pain, and continuous vomiting from her kidney infection, the anxiety in her voice was palpable.

"Mom gets real frightened when she's alone at night—it's the Alzheimer's, you know," Rita had fretted to Pascal that first morning of hospitalization. "The neighbor lady across the hall stayed with her last night, but she don't know if she can do it tonight." Pascal has heard this kind of concern many times before, in one form or another: the person afflicted with AIDS worries more about a loved one—a parent, a child, a lover, a pet—than about his or her own condition.

"I tried to hold off on coming to the hospital, but the pain got so bad," she apologetically explained. "Mom got real scared the sicker I got, right before the neighbor called 911 for us," she said with an added desperation that stemmed from an even greater burden.

"You see, Sister, Mom don't know I'm HIV," Rita confided. "It would *kill* her to know. I'm all she's got. If anything happened to me . . ." Rita stopped midsentence, not wanting to break down in front of Pascal.

Rita had not yet suffered any complications from her HIV infection, except for minor problems such as thrush and shingles, annoyances that could easily be hidden from an eighty-one-year-old mother stricken by severe arthritis and early Alzheimer's. But over the past two years, despite faithfully following her clinic doctor's advice, Rita has seen her T cells steadily fall to the 150 range, a count that has earned her an AIDS diagnosis, with all the risks of major complications therein.

"I *have* to stay well, Sister Pascal," continued Rita a week ago. "Every day I pray to the Lord to take Mom before He takes me. The Lord is all we have. With Him I can bear all things." The tears were streaming down her face, but there was no sobbing or wailing. "Yes, Sister, with the Lord, all things are . . . possible."

"Tell me," Pascal quickly interrupted, both to help Rita compose and to learn more about her situation, "what church do you and your mother belong to?" Rita seemed to appreciate the chance to change the subject.

"The First Church of God, but it's been ages since Mom could go—her arthritis acts up so. I try to make it every Sunday, but sometimes I can't when I have to work at the hotel." A part-time maid at a Midtown luxury hotel, where the charge for a regular room can exceed $350 for one night, Rita has never received full health benefits and has thus ended up at St. Clare's.

"Maybe the pastor in your church could help us with your mom," suggested Pascal. "Our social worker could maybe get a home attendant to watch her while you're here."

"I don't mean no ingratitude, Sister, but it's more than that. The least little time I'm gone, Mom worries so. Sometimes when I come back, she's sitting there crying. When I ask what happened, she just says, 'I miss you, don't leave me,' over and over again. I worry how she's going to take all this."

"But what if we got you a phone right here in your room, so you could talk with your mom whenever you two wanted?" Always tenacious, Pascal was not about to let Rita's anguish increase from lack of a telephone, which is not a standard amenity in Spellman rooms and must be rented for seven dollars a day.

"Oh, Sister Pascal, that would be *wonderful,* but we don't have a phone at home, not since Dad died." Rita sounded a little amazed that anyone would think they had such a luxury as a telephone in their apartment.

"Well, Rita," concluded Pascal, "we'll have to take care of that, won't we?"

Indeed, Pascal was able to do something few medical workers would take the time to do. She and Rita's social worker arranged for not only a home attendant but also, owing to a sympathetic NYNEX supervisor, a temporary emergency telephone for her mother. Pascal's "slush fund"—her petty-cash fund for good works—financed a phone in Rita's hospital room, and Rita's pastor was able to see to it that her mother had three meals a day. Pascal's telephone conversation with their pastor, the Reverend Mr. Stoner, confirmed her belief in Rita.

"Sister Pascal," replied a stiffly formal, but obviously concerned, Reverend Mr. Stoner, "Sister Rita has been a pillar of our church. She has given so much of herself to the Lord's work, but she has always been somewhat reluctant to let us return the help she has given so many of our members. It matters not to us why she is in the hospital." Pascal, of course, did not divulge anything about Rita's illness, but AIDS was certainly no stranger to his congregation. The reverend continued, "Jesus cared only about our souls' relationship with him. Thank you so much for letting us know about Sister Rita's needs. Our church and our Lord will see to them."

The remainder of Rita's hospital stay was one of rapidly diminishing worries. Under the constant care of the home attendant, her mother was reassured by the telephone link, and Rita's peace of mind allowed her to stay for the full course of antibiotic therapy for her kidney infection.

Now, a rejuvenated Rita B. is exuberantly hugging an equally delighted Pascal.

"God bless you," Rita shouts out again. "You are an angel of the Lord, and I will pray for you and your work every day of my life. And may the Lord give you a merry Christmas."

"And I will pray for you and your mother to have a blessed New Year!" Pascal says, and kisses her "old friend" farewell.

Often, pastoral care has far greater impact on a patient's family or friends than on the patient, as is illustrated by Pascal's next visit today, which, like many of her encounters, is completely impromptu. A Spellman unit's head nurse—a serious, straitlaced woman in her forties—quietly approaches Pascal with a delicate problem, which she relates in hushed tones.

Thomas Q., a twenty-eight-year-old man with recurrent kidney stones, was admitted last week to the unit's only four-bedded room and has apparently been keeping his three roommates up late the past few nights by having sex—fairly loud and raucous sex, if reports are to be believed—with his young girlfriend. A muscular bodybuilder with T cells of 1,200, Thomas is probably the healthiest patient on the Spellman service right now. He is also one of the most stubborn and sullen, never taking well to people telling him how to behave.

"When I told him about the complaints, he informed me, 'I can kiss my wife *any time, any place, any way* I like, and there's nothing you or anyone else can do about it.' "

The nurse is obviously perturbed, and embarrassed, that she has to referee such behavior. "Sometimes I feel I'm in a nursery school instead of a hospital! I mean, no sooner do I step into that room this morning than the other three patients are on my case about the noise last night. They said they didn't even close the privacy curtains when they were doing it. Talk about disgusting!"

Trying hard to appear concerned, Pascal listens with mock solemnity before replying.

"As a great philosopher once said, sex is fine until it disturbs the neighbors in the next county, so I suppose Thomas's escapades fall into that category," she quips, quickly realizing that the frustrated nurse wants more than a glib retort. "Let's see what I can do," Pascal adds, assuming a more serious air.

Sex on the Spellman service is, of course, not as rare as a UFO encounter. Whether between patients or between staff members or, very infrequently, between a staff member and a patient, the stories are salacious enough to shock the prurient of heart. But Thomas Q.'s nocturnal antics may have elevated "the wild thing" at St. Clare's to new heights of impropriety.

After dutifully knocking and allowing a decent interval to pass, Pascal carefully enters Thomas's room, loudly announces herself, and cautiously peers around the privacy curtain drawn around his bed. His "wife," Iris—actually his girlfriend—is there alone, perched on the edge of the bed, snapping her chewing gum and primping in a handheld cosmetic mirror.

"He's down getting an X ray," she says absentmindedly, not taking her eyes off the mirror and the front bangs she is teasing with a comb. In fact, all of the other three roommates are out, too, probably to smoke in the nearby patient lounge.

"That's all right. Actually, I was really hoping to talk with *you,* since we haven't met before," Pascal gently intones, hand outstretched in greeting.

Iris is taken aback. She seems surprised anyone would want to talk with just her. This older woman's manner seems so sincere and friendly, not at all like that of other hospital staff who have been giving Thomas so much grief about their conjugal groans. Putting her comb and vanity mirror down on the bed, she stiffly shakes Pascal's hand and nervously straightens out the red miniskirt tightly covering the upper half of her thighs. Despite the heavy makeup, Iris still looks like a little girl trying too hard to be an adult.

Pascal introduces herself and asks how Thomas has been doing. Nodding satisfaction at Iris's assurances that he is "just fine"—a foregone conclusion, given his prodigious sexual gymnastics from the night before—Pascal moves to the subject at hand.

"Iris, the nurses tell me you and Thomas kept everyone up last night." Pascal's tone is neither accusatory nor flippant. Iris's posture instinctively straightens up with these words, but before she can protest, Pascal tries to reassure her, with words Iris has probably not heard from a nun before.

"I think it's just wonderful that Thomas is feeling so well he can make love all night long. I really mean that, but, you know, Iris, the other patients need their rest. They're really sicker than you might think."

"We didn't make *that much* noise," Iris replies petulantly, trying hard to act both indignant and mature, despite her appearance, which suggests she is about sixteen. "Besides, it's no one's business but ours. Thomas says we can make love anytime we want. Anyway, wha—wha—what would *you* know about lovemaking anyway?" Iris's voice trails off ever so tentatively as she hurls this latter challenge at Pascal. Iris is not quite sure she really wants to offend an older nun, especially one who seems to be friendly enough.

Rather than taking offense, Pascal seems to welcome Iris's retort and chuckles good-naturedly. "Iris, my dear, I wasn't born a nun, you know. I was once your age and had many of the feelings you have now. Why, difficult as it might be for a young lady like you to believe"—and here there is neither rancor nor sarcasm—"I might even *still* have such feelings. And as for men, let me tell you, I'm an old Italian from way back, and believe you me, that means I know how men can be." As she chats with Iris, Pascal radiates good humor. She leans over and gently taps Iris on the arm, as if to say, "We're both women, and as women, we're in this thing called life together."

With Pascal's amicable touch, Iris breaks into a small, self-conscious smile. She seems especially self-conscious, even a little uncomfortable, when Pascal remarks about "how men can be."

"I know what we did wasn't just right"—Iris's tone is now more conciliatory—"but, Sister, he's all I've got, and I've got to keep him happy."

"And so you should, just as the good Lord intended, but maybe you should think about your own pride as a young, attractive woman. You see, Iris, you can have wonderful sex with Thomas and still respect yourself. You can still respect your dignity as a woman—a woman who loves her man *and* loves herself."

Pascal waits a few seconds for her words to sink in before concluding, "You see, Iris, we really can't love anyone until we love ourselves first."

Pascal has gotten Iris to thinking, thinking about herself and her dignity—all concepts largely foreign to her.

"Maybe you should talk with Thomas about this," suggests Pascal. "You could start by telling him your love is too special to be put on public display before strangers. Tell him Sister Pascal thinks you two deserve better than that." Pascal has looked in on Thomas Q. during his prior admissions for kidney stones, and she feels their previously cordial visits will lend some weight to her present advice.

Still pondering Pascal's words, Iris seems subdued, even embarrassed. "My mother never talked much about these sorts of things. I guess . . . I guess I still have a lot to learn."

"We *all* do, Iris. And if ever you want more down-to-earth girl talk with an old gal who's seen it all, you can give me a call anytime." Shaking hands again as she gets up to leave, Pascal delivers the coup de grâce: "Remember, Iris, you are a very special, very beautiful young woman, and don't ever let anyone tell you—or treat you—different!"

After noon mass and a light lunch in the cafeteria, Pascal marches off to keep an appointment she made with Carlos R., whose brother Luis is an inpatient on Spellman. Carlos is taking time off from work to see Pascal today, so she wants to be as prompt as possible. In conversations with concerned family members such as Carlos, pastoral care workers sometimes uncover previously unappreciated emotional anguish that merits as much and often greater attention than that of their sick loved one with AIDS. Although Pascal and Carlos have never before met, he is about to reveal to her a desolation that he has not even discussed with his wife, let alone other Spellman staff.

Many Spellman patients are visited by concerned mothers and sisters, but the fathers and brothers are much less frequent visitors on the wards, often because these male relatives are likewise sick from AIDS, are in prison, have abandoned their families, or are dead. Luis R. is somewhat unusual in having a concerned older brother, Carlos, who visits him daily, their mother being too frail to visit regularly and their father having left them many years ago. In the advanced stages of AIDS—his T cells are 8—Luis is a twenty-eight-year-old ex-prisoner who is recovering from a brain abscess, an infected pocket of pus in the brain, which he got from shooting up with cocaine. Despite his drug abuse and past problems with the law, Luis is genuinely affectionate. He has been a pleasant, cooperative patient who never complains. Both the brain infection and HIV dementia have progressively limited his ability to get around, and he has often needed help eating and getting dressed. Strapped as they always are for time, Luis's nurses have done their best to assist him, but his brother's continual help has been invaluable in keeping him functional: because of Carlos, Luis is not bedridden.

Almost every evening, after he finishes his job on a construction crew in Brooklyn, Carlos makes the trip into Manhattan to see Luis, spending up to two hours helping to feed him, walking him in the hallway, bathing him, and rubbing his back with skin lotions to relieve the irritating rash many HIV-infected people have. On weekends, their mother and Carlos's wife and older daughter also visit, but it has been Carlos's ministrations that have gratified the staff most. Both the evening nurses and the Spellman social worker, who has talked with him over the phone, have repeatedly commented on Carlos's concern for his younger brother, such fraternal devotion being distinctly uncommon on the Spellman service. On the surface, Carlos does not appear any more beatific than legions of other burly, hard-hat workers, but the image of him gently massaging his brother's back or patiently feeding him dinner has added a whole new dimension to Carlos's typically blue-collar appearance.

Everyone—Carlos, his mother, and the medical team—agrees that Luis will require nursing home care, but the location of the nursing home seems to have become a source of contention between Carlos and the Spellman staff. Because he and his family live in Newark, New Jersey, Carlos has been insisting that Luis be placed in a nursing home there. However, Luis's welfare benefits entitle him only to care in New York, and the resultant conflict has threatened to derail attempts to get him into the long-term care AIDS unit at Terence Cardinal Cooke Hospice. Never unpleasant, but firmly adamant in his wishes, Carlos recently announced that he would take Luis to his home in Newark rather than agree to transferring him to "that place," as he would refer to Cardinal Cooke. Such a development would be unfortunate, since Luis would then be without needed home health care and might deteriorate faster. Pascal has never had a chance to talk with Carlos, although she has often stopped in to see Luis, who, because of his dementia, really cannot remember much from one visit to the next. Trying to resolve the conflict over Luis's placement, his social worker has asked Pascal to intercede.

After briefly exchanging pleasantries at Luis's bedside—as always, Luis does not say much but is always gratified to have visitors—Pascal and Carlos step out into the hallway to talk. Immediately, Carlos apologizes to Pascal for, in his words, "dragging" her into the dispute about Luis's nursing home placement.

"It's just that we want him closer to us. He's all the family that's left, since my two younger brothers died in Sing Sing last year." Pascal does not need to ask about the causes of their deaths. In New York of the 1990s, male Hispanic state prisoners die of only one thing.

"Carlos, the way I see it, you need to do what you think is best for you and Luis. If that's moving him to Newark, then so be it. Are you thinking about your place or your mother's?" As always, Pascal speaks without guile, realizing that patients' families have needs, too, even if those needs run counter to ideal social service planning.

"I guess it would have to be my house. Mom is too sick to have Luis there. To tell the truth, Sister, I just want what's best for Luis, even if it isn't Newark. I'd like him close to home, but I'm not sure how we'll do that." Carlos's voice is trembling slightly, an indication he has not really thought through the idea of taking Luis home.

"All I do know," he continues, his voice more composed, "is that I don't want him at Cardinal Cooke."

"Is that so?" replies Pascal, her question inviting Carlos to elaborate on his concerns. Something in Carlos's voice tells her that his concerns are much deeper than worries about Cardinal Cooke.

"No question about it. That place is in a slum"—Terence Cardinal Cooke Hospice is at Fifth Avenue and 106th Street in Manhattan—"and I wouldn't want to take my mother there to even visit. It's way too dangerous."

"Really? Have you been up in that neighborhood before?"

"No, but it *has* to be in a bad area. I mean, it's all the way up at One Hundred and Sixth Street." Carlos speaks with certainty, despite the fact that, like many people from nearby New Jersey, he really does not know much about various New York neighborhoods. Any place, he contends, that far north of the invisible border of Ninety-sixth Street cannot possibly be safe.

Pascal can barely conceal a smile. She has been to Terence Cardinal Cooke several times and knows it is not only safe and secure, but also a gracious residence, having been built into what was once a luxury apartment building. Although Fifth Avenue at 106th Street is, strictly speaking, just to the north of the Gold Coast and is not prime real estate, this location is hardly a "slum."

"Tell me, Carlos—off the record—what do you think about the care Luis has received here at St. Clare's?" Pascal knows that the amenities at Terence Cardinal Cooke are equal to or better than those at St. Clare's.

"Everyone here's been great, Sister. This is where my brothers came before they died, so I know this place pretty well. I mean, there *are* problems. There sometimes aren't enough nurses, the toilet sometimes leaks and all that, but it's okay here. Luis likes it, and everyone here is nice to him, so that's all I need to know."

"Well, Carlos, let me tell you"—here Pascal speaks with a frankness Carlos could never doubt—"if you think St. Clare's is okay, you'll *love* Terence Cardinal Cooke. It's much more spacious, and the people there are every bit as nice as anyone here—*and,* there are more of them, so Luis won't be overlooked when sicker patients take up the nurses' time, like sometimes happens here. The locale is as safe—*safer*—than this neighborhood. I think if you were to visit Cardinal Cooke, you'd be pleasantly surprised."

Carlos listens quietly, but without eye contact, as Pascal speaks. He is staring out into space, appearing troubled and preoccupied—not by what she is saying, but by much larger matters.

There is a brief pause after Pascal's last words. Arousing himself out of his reverie, Carlos looks at Pascal. He appears unspeakably sad. Tears begin to form in his eyes.

"No, I haven't seen Cardinal Cooke, I know I should visit . . ." Now he is beginning to tremble.

"Sister, I don't know what to do . . . first there's Sammy, and then Jose, and now . . . and now Luis . . ." Carlos can barely get out the words, quietly sniffing back the tears. Instinctively, protectively, Pascal gently guides him into a nearby corner and shields him from the passersby in the crowded ward hallway. As Carlos wipes his eyes and blows his nose with his handkerchief, Pascal softly pats his shoulder with one hand and holds her other hand against her chest, clutching the silver cross hanging from her neck. She says nothing, but both her touch and the gentle swaying of her upper body convey an infinite compassion and concern.

"I don't want to lose Luis. He's all that's left. It'll kill Mom—she hasn't been the same since Jose died." Carlos's outburst is now under better control, but the grief and worry on his face are even more intensified. "It's not the nursing home business. I don't want him that far away from home, but if it has to be, it has to be. But I can't go through another funeral, Sister, *I just can't.*"

It is an incongruous sight, the two of them huddled together in a hallway corner as the busy ward traffic whizzes past. Here is this tall, powerfully built construction worker, unraveling before an older nun. Although she has done this many thousands of times at St. Clare's, Pascal never appears coolly composed and detached as professional "bereavement counselors" often do. Her manner with Carlos, as with all others, displays an uncanny strength, mixed with tentativeness and humility.

"Carlos," Pascal says, breaking her silence, "with love comes strength. God gives it to us when we need it the most and when we expect it the least. Your love for your brothers Sammy and Jose has given you the strength to care for Luis and your mother. And your love for your mother and Luis will give you strength not only to survive but to love even more." Pascal's declarations instantly seize Carlos's attention. There is now rapt eye contact between the two of them, and her final summation seems incontestable: "Your gift for love comes from God, and take it from me"—here Pascal taps him on the arm and winks at him—"God is not going to leave you out on a limb after all the wonderful things you've done for your brothers. You have my word on it."

Before they part, the two of them trade phone numbers. She promises to call him in a few days to see how he's doing and urges him to call her anytime he has worries about Luis.

Carlos and Pascal return to Luis's bedside, for further brief pleasantries, which are shortly interrupted by a beeper page for Pascal to a nearby Spellman unit. Assuring Luis she will stop by to say hello tomorrow, Pascal leaves the two brothers to return the call from the page, which is from me, apprising her of a pending death.

"Pascal, it's Baxter. Sorry to bother, but Felipe A.'s worse today. I can't be sure, but I think he's gonna check out soon, maybe tonight, but then again, he's fooled us before." Pascal appreciates such alerts, even if they turn out to be false alarms.

"Oh my, is that so?" A note of cheerful expectation is in Pascal's voice, indicative of her hope for Felipe's release from this life. "Thanks a million, Dan—I'll be by in a bit." Felipe A. is well known to Pascal, who, unlikely as it may seem, regards him as a *very* important patient.

However, Felipe's impending demise must wait, for no sooner does Pascal hang up from my call than her beeper goes off again, this time with even more serious news. A code has just commenced on the Spellman floor directly above her. Pascal rushes up the stairs to see who is being coded and what, if anything, she can do to help.

Pastoral care is always STAT-paged whenever emergency cardiopulmonary resuscitation is initiated, so that Last Rites can be administered and emotional support rendered to any distraught family hovering outside the room where their loved one is being pounded and pumped on. For pastoral workers such as Pascal and Father Jack, a code usually signifies apocalyptic announcement of a soul's transition from this world to another—the success rate of in-hospital resuscitation attempts is a dismal 10 percent, or less. Given the typical havoc of a code, as well as American medicine's a priori denial of existence of the soul, the presence of pastoral care staff at a code is always somewhat anomalous.

This afternoon's CPR is on Sarah W. and comes, like most codes, without warning. However, in a certain sense, Sarah's code was bound to happen.

A twenty-six-year-old Dominican female with several major complications of AIDS—right-eye blindness from CMV retinitis has been her most recent—Sarah has never been able to stay away from cocaine. Her frequent hospitalizations have essentially been continuous crack parties, with medical care more of an afterthought for her. More childlike than her six-year-old son, who is in a foster home and sometimes visits her at St. Clare's, Sarah has often disappeared for hours at a time from her hospital room, to be found smoking crack in a stairwell or on a fire escape. She created a stir a few days ago when, at 2 A.M., the hospital's security cameras spied her having oral-genital sex with a male patient. They were just outside an emergency exit door in five-degree weather, unaware they were in full view of the hidden camera overhead. This escapade almost got her kicked out of the hospital, but because she is homeless, it was decided to put Sarah on one-to-one observation, with a nursing aide watching her at all times.

Unfortunately, the aide assigned to Sarah today was slightly under five feet tall and weighed in at 310 pounds—her girth was an ample sixty inches. Predictably, Sarah was easily able to outrun her chaperon, once again disappearing from her floor earlier this morning.

Just a few minutes ago, a nurse found Sarah unresponsive, without vital signs, in a linen closet on the Spellman floor above hers. A warm crack pipe was nestled beside her. In rapid succession, a code was announced via beepers and overhead page, and the code cart was wheeled from down the hallway. After dragging Sarah's limp body out of the closet and into the hallway, the gathered host of Spellman staff is starting CPR, right on the floor next to the linen closet. By the time Pascal reaches the unit, the hallway is clogged—it is a gridlock of code cart, oxygen tank, IV poles, and the dozen or so people huddling and kneeling around Sarah's thin, ninety-pound body, which is being pumped and pounded.

As the ampoules of epinephrine are effortlessly pushed through the IV line, and as the chest is compressed with regularity, there is little conversation among the staff, all of whom knew Sarah well—her exploits over the years were legendary throughout the Spellman service. Most codes, especially those in such cramped, unusual places, are bedlam, but Sarah's is strangely calm, perhaps because of the incongruous proximity of holiday garlands and tinsel decorating the corridor. In the businesslike silence of this code, the universal, but unspoken, feeling is that this pathetic ending was inevitable—that, given Sarah's insatiableness for drugs, this routine of obligatory chest pumping and electroshocks was bound to happen.

Pascal keeps the curious onlookers—mostly other patients and their families—at a safe remove. There is no need to be on the lookout for any of Sarah's family stumbling onto her final drug high, since, except for her young son, she has no one. Arriving a few minutes after Pascal, Father Jack stands behind the PA who is on his knees, vigorously pumping on Sarah's chest. Like everyone else, Jack shows no emotion, no surprise, and utters a silent benediction for Sarah, who never professed any religious inclinations during her many stays at St. Clare's.

Proceeding on the dusty floor of the hallway, Sarah's CPR this afternoon will be extraordinary not only for its venue. For, as the minutes tick by, it is becoming evident that Sarah, like a wraith that refuses to disappear completely, is literally slipping into and then out of death. When, after ten minutes of CPR, it appears her heart will not start, and when the assembled medical staff conclude there is no hope and accordingly stop resuscitating her, her heartbeat and breathing spontaneously reappear—but just for half a minute or so, only to peter out and stop, thereupon obligating the doctors to resume pumping on her chest, to try to revive her again. When another ten minutes of CPR is futile in restarting her heart, the doctors again stop the pumping and agree it is time to declare her dead—but just as they are about to cover her up, Sarah's heartbeat and feeble respirations once again restart, only to stop a minute later, requiring yet another round of chest pumping. The procedure becomes grotesque—there is simply no other word. Sarah is oscillating between life and death, this macabre sequence of events occurring two more times, to the consternation of all the staff circled around her motionless body, a body that seems unable to live without CPR or to die promptly if left unattended.

After the fourth revival and subsequent deterioriation of Sarah's vital signs, one of the frustrated doctors half-jokingly turns to Father Jack, quasi-urgently imploring, *"Father, do something!"* Whereupon, Jack, who is likewise exasperated by this repetitive miniseries of death and resurrection, looks up to the heavens and, with arms outstretched over Sarah's body on the floor below, solemnly admonishes God in a loud and impatient voice, "Lord, *make up your mind!* Either take her or leave her with us!" No one seems too surprised by Jack's provocative command to God. As both comic relief and desperate petition, it seems perfectly congruous under the circumstances.

Called on the carpet by His no-nonsense priest, the Lord finally takes Sarah after the fifth round of CPR. But her doctors warily watch over the shade of Sarah for several minutes before declaring her once-and-for-all dead, the extra delay only prolonging the congestion in the hallway.

As the nurses are stuffing Sarah's corpse into a body bag on a gurney, Pascal and Jack confer about how Sarah's son should be notified, finally agreeing that the child welfare worker should coordinate the matter with the foster family. Somehow, the yuletide death of a six-year-old boy's mother, his only family, seems even more difficult to contemplate than the gruesome spectacle just concluded on the unit's dirty floor.

"Sarah sure went out with a bang, didn't she," Jack quips to Pascal. "She kept us guessing up to the very end where she was—in heaven or still here with us. Just like when she was always running off to do crack."

The unmistakable affection in Jack's words, and in Pascal's smile of agreement, indicates that this levity is not at Sarah's expense, but rather is directed at the pretensions of medical science, which is often uncertain when death commences, despite the frantic, usually futile, codes to stave off this perceived adversary.

Sarah's CPR has set Pascal further behind schedule, and checking her list of patients, she determines, like the doctors and nurses she works with, who can safely be put off until tomorrow and who must be seen today. Because she postponed Todd S. from yesterday's hectic rounds, she definitely must look in on him today.

Spellman physicians often turn to pastoral care when a patient refuses to cooperate with what the staff deems optimal medical care. Just as Pascal's peacekeeping mission earlier today defused a prickly situation with a histrionic Georgie J., so has Todd S.'s ward team asked her to gain his acquiescence to a Groshong catheter. Most of the time, the medical staff is too busy, or uninterested, to understand the real reason for a patient's recalcitrance. They simply want pastoral care to extract the patient's willingness for this or that test or procedure. Pascal, of course, is neither a magician nor a "body mechanic" for the soul. So often in difficult situations, the nun, rabbi, imam, or priest serves as a confessor, listening to secrets and confidential issues that the patient cannot share even with the doctor. Such was the case with Todd S.

Todd S. is having another flare-up of thrush in his esophagus. The white fungal infection is coating his gullet so heavily that he has been unable to eat solid food for two weeks, and over the past few days just swallowing liquids has become torture. Todd definitely needs another course of intravenous amphotericin B to clear up this intractable thrush; it will be his fourth go-round over the past six months with this same problem. Moreover, Todd's doctor is advising that he get a permanent Groshong catheter—a large intravenous line inserted into his upper chest—so he can receive long-term amphotericin therapy, the hope being that he will eventually be able to eat solids again and regain some of the thirty pounds lost during the past weeks. Chronic amphotericin treatment, which Todd could easily self-administer at home, would obviate the frequent admissions and wide swings in weight he has had over the past half year.

But Todd has been refusing to have this central IV line put in. Whenever the time for the procedure would arrive, he would have some excuse—either he is "too tired" or is "too sick" or just does not "feel like it." Efforts to beg and cajole him have been fruitless, as has been counseling by the psychiatrist and social worker. Todd is severely depressed: over the past two weeks, quiet resignation has supplanted his normally abrasive personality. To the frustration of his ward team, Todd seems to have given up way too soon, for, although malnourished, he would probably stabilize on long-term amphotericin, eventually improving enough to be able to go home and resume his life, *if* he allowed the Groshong catheter.

Yesterday, Todd began to refuse all his medications, and—the only hopeful development—he started up again with his "difficult" personality, but with added vengeance. Loudly cursing anyone who tries to give him his pills, to take his vital signs, or even to change his bedsheets, Todd is rapidly becoming unmanageable, and no one can understand why.

Pascal has looked in on Todd several times since his admission two weeks ago, but so far she has not been able to connect with him on anything other than a superficial level, their conversations dwelling solely on pleasant generalities. Always courteous, Todd would nonetheless maintain his distance, never confiding his feelings. As with Father Raphael B. earlier today, Pascal senses in Todd's room a desolation of the spirit, but with a hopeful caveat: whereas Raphael's clerical past has probably consigned him to perpetual silence, Todd has never suffered from guilt, or even reticence, about his private life.

Indeed, on first glance, Todd is very different from Pascal. Born and raised in an impoverished "holler" of rural West Virginia, where his childhood interest in dolls earned him the nickname "the doll boy," Todd ran away from home at age sixteen. His father, a Pentecostal minister, could never accept Todd's all-consuming desire to be a woman. After hitchhiking to New York to have a sex-change operation, Todd initially hustled the old men and tourists trolling Times Square's porno movie houses, eventually saving enough money to get large silicone breast implants done by a cut-rate plastic surgeon in Staten Island. The results of cosmetic surgery were exceptionally bodacious, propelling Todd, who had always dreamed of being a self-styled "male actress," into a new career as a featured showgirl at one of Fire Island's notorious cabarets. In backhanded tribute to his West Virginia origins, he assumed the well-worn stage name of Iona Trailer. Indeed, by then Todd felt he had reached a certain level of respectability within his social circle, at least removed from the more dangerous street world inhabited by Georgie J. Unfortunately for Todd, with the money there also came cocaine, ecstasy, and all-night orgies. By the time he was twenty-four, he had finally saved enough money for "the real thing," as he always called a definitive sex-change operation, but pre-op labs revealed several abnormalities that caused him to get HIV-tested for the first time. The subsequent positive results made the surgeons reluctant to effect "the real thing" for Todd.

A month after testing positive, Todd began having severe chest pains whenever he ate, and a diagnosis of *Candida* esophagitis was eventually made. Standard outpatient medications did not relieve his worsening symptoms, and intravenous amphotericin B was the only treatment that eradicated his pain.

Today, Todd almost seems to be expecting Pascal's visit.

"I suppose you're going to tell me I should get the fucking Groshong catheter, too?" he sarcastically demands as soon as he sees her at the doorway, as if relishing yet another confrontation with the staff.

"Oh, no, not really." Pascal shrugs, her eyes turned upward to the ceiling. "I just thought I'd stop by and say hello." Pascal's studied disinterest does not make her an inviting enough target for any invective, much as Todd would like to lambaste anyone who dares come in. Her I'm-just-an-innocent-bystander-so-please-don't-yell-at-me pose reassures him that she has no agenda and is different from everyone else visiting him recently. Warily, he nods for her to come on into his room.

"I'm sorry I used the F-word, Sister, but everyone here's been in my face. They keep telling me I need to take my medicines, I need to do this, I need to do that. They're telling me I have 'everything to live for,' that I'm 'not sick enough to die yet,' and all that other bullshit. Has it ever occurred to them I don't give a dead rat's fart what they think, Sister? Has it ever occurred to them maybe I *want* to die?"

Like a little girl standing before a stern headmaster and trying hard to be on her best behavior, Pascal stands erect at the bedside, hands tightly clasped and folded in front of her. Nodding her understanding ever so slightly, she says nothing.

"Nobody understands, nobody understands," Todd mutters out loud to himself, his voice drifting off.

A brief pause ensues, then Pascal, still nodding her agreement, softly replies, "Yes, Todd, you may be right about that, you may be right . . . maybe nobody can understand." Although Pascal still sounds noncommittal, her voice cannot disguise a mantle of basic warmth.

"Yeah, you're right: nobody understands. They just don't know how many years I've scrimped and saved to buy these beautiful tits, Sister. They don't care that they're all I've got, they're what makes me *me,* and now they want to take it all away with a goddamned Groshong." Lying in bed with his hospital gown partially open at the top to expose his bare chest, Todd wistfully gazes down at his breasts. They are indeed striking in their size and perfect symmetry, both of them like firm melons, silicone orbs almost too round and too perfect to be real. Todd is tall and lean, with thin, straggly blond hair down to his shoulders. Although not overly hirsute, a copse of hair is on his chest, including his breasts.

"Nobody understands," he laments, staring forlornly at his breasts. "All I want is a few more days to enjoy my beautiful, big hairy tits . . . just leave me alone to enjoy my big hairy tits, and then I'll let them put the goddamned catheter in my chest." Although indulging in theatrics, Todd caresses his words *hairy tits* with both humor and seriousness.

"Do you understand what I'm saying, Sister? Does any of this make any sense to you?" Todd's query is part challenge and part plea that perhaps Pascal might actually care what he is talking about.

"Well, Todd," Pascal tentatively replies, as if she is sincerely trying to learn something from him, to put herself in his place, "I think I understand that you really like your hairy breasts, and that these breasts mean a lot to you. I understand you feel the Groshong will damage your breasts or, at a minimum, make you feel less attractive, less of the woman you'd like to be."

Todd instantly looks up at Pascal, almost with disbelief, at her words—"less of the woman you'd like to be"—words that signaled an empathy, an acceptance no one has obliged him so far. Everyone else trying to convince him to agree to the Groshong painfully ignores his silicone breasts.

"And you know, Todd"—and here Pascal leans forward and taps him on the shoulder—"you know, you really *do* have gorgeous tits, if I say so myself."

"You really think so, Sister?"

"You have my word on it, Todd."

For the next five minutes, Todd and Pascal chat about his breasts—when he got them, who did it, how much it cost, why he decided to have them exactly alike rather than slightly asymmetrical, and how many cabaret jobs they got him over a brief span of years.

"It took me two years to save up for my breast surgery," he eagerly relates to Pascal, "and rather than wait till I had enough for both my tits and my sex-change operation, I got the tits first. I'd joke to people that my pussy was 'on layaway at K Mart.'" Todd is not trying to scandalize his older visitor—he is just being himself.

Todd holds forth on Fire Island's pecking order of drag queens, partial transsexuals, and completed transsexuals, until he is tired, his eyelids drooping. Knowing when to make her exit, Pascal promises to come back tomorrow.

"I'd like to hear more, later," she concludes.

There are no further explanations as Pascal leaves Todd's room. More than any doctor or nurse, she has begun to secure Todd's trust. There will be opportunity for them to talk about the Groshong catheter tomorrow and the day after, when she will assure him that his breasts will be left largely untouched by the Groshong. Todd, in turn, will unload onto his new friend his *real* anxiety—his secret worry about what will happen when his father, who is visiting for the first time right after Christmas, learns his son not only has big, hairy breasts but also has AIDS.

Before seeing the next patient, Pascal first has to pick up a package and makes a quick detour to her small office, on the top floor of the oldest part of the hospital, where narrow hallways and cramped cubicles are all that remain of what had once been the private quarters of the first nuns who tended St. Clare's two generations ago. The package Pascal fetches is a surprise for Doris J. It will be the only Christmas gift for Doris's three small children, who are living in a foster home while she is in the hospital.

Doris was admitted a week ago for bacterial pneumonia and a flare-up of her asthma, both chronically recurring problems that have landed her in the hospital almost every other month over the past two years. A congenial but nervous lady—a woman whose forty-two years of life are severely etched by the lines of her face—Doris has always hung onto the edge of life, both socioeconomically and emotionally. As if her life as a poor single parent were not complicated enough, AIDS is now competing with all the other myriad worries tormenting her. Indeed, the litany of worries seems interminable: stretching the food stamps to the end of the month, medical care for her children, crime in their neighborhood, making the monthly rent on their two-room Harlem apartment, hoping the landlord will finally turn on the heat this winter—these and a dozen or so other distractions always preoccupy her as she lies in her hospital room. Moreover, Doris must face the constant lure of drugs, which still periodically ensnare her, despite her best attempts to resist them. Indeed, as tribute to intensive intervention by the staff in the methadone clinic, Doris has been drug-free for the past nine months, a not insubstantial triumph, given the multiple battles she has been fighting.

Nonetheless, the holiday season is fraught with gloom. Because her current hospitalization makes her ineligible for a welfare check this month, there will be no gifts for her three daughters, ages six, seven, and nine.

"I wish I could do something for them. You never know, this might be our last Christmas together," she told Pascal a few days ago. Never acting entitled—she could never be cited in a campaigning politician's sound bite against welfare mothers—Doris has always been a grateful patient, and despite her history of drug use and brief incarcerations for minor offenses, she has always been a caring mother who, like most other poor mothers in the city, has never neglected her children, even when she was sick in the hospital.

"I may have made a lot of bad mistakes in my life," she recently told Pascal, "but my children ain't one of 'em. They're the only things keeping me going. The good Lord and my book will see to it they know who their mama was when I'm gone."

Doris's "book" is both a personal journal and a collection of letters she has been writing to her daughters over the past six months. The dog-eared brown spiral notebook is always with her, a testament of courageous love she hopes will someday speak to her children from the grave.

Today, Doris is stunned as Pascal, bubbling like one of Santa's elves, drops off her surprise, a shopping bag containing three identical gifts for each of the three children—a winter scarf and glove set, a large box of animal crackers, and a Barbie doll complete with an extra fashion outfit.

"I figure you can't go wrong with Barbie dolls for little girls," Pascal says. She motions over to the framed photo of Doris's daughters that is standing on the bedside table. "I thought you might like to wrap them yourself—there's scissors, tape, and paper at the bottom of the bag."

Before the tears can even form, before Doris can even try to blurt out her dumbfounded thanks, Pascal has already turned on her heels and is halfway out the door. Pausing to look back at the photo on the bedstand, Pascal declares, "You're a great mama to those beautiful daughters of yours. I'll stop by tomorrow to see how we can get these to them by Christmas."

The day is almost over. The snow-laden mid-December sky is already darkening over the tenements of Hell's Kitchen. Pascal sets about taking care of loose ends.

First is Romeo L., a thirty-one-year-old patient from the Dominican Republic, who received notice a few hours ago from his parole officer that he will be deported tomorrow morning as an illegal alien—never mind that he is undergoing sight-saving therapy for CMV retinitis. Sometimes pastoral care petitions carry more weight than a physician's or social worker's. Thus, at the request of Abby A., Pascal faxes off an urgent appeal to the Immigration and Naturalization Service, and for good measure, she calls her friend in the archdiocesan office for help. This friend knows someone in the Justice Department who, as a last resort, can pull strings for Romeo, at least until the initial phase of his anti-CMV therapy is completed.

Then there are the baptisms scheduled for two Spellman patients tomorrow afternoon, in addition to what Georgie J., Pascal's "prettiest hooker in Times Square," will proudly regard as her fifth baptism. Father Jack will actually administer to her the Sacrament of the Sick, perhaps with an innocent splash of holy water on the side. Pascal quickly stops by the chapel's sacristy to make sure the baptismal certificates and requisite liturgical supplies are on hand for the ceremony. She plans to visit the other two baptismal candidates tomorrow morning, to go over the order of the service and to give them a spiritual booklet to review. Pascal knowingly concludes that Georgie J. is already an experienced hand at Catholic baptisms, having been baptized once and anointed with holy oil three subsequent times at St. Clare's alone. As Pascal routinely tells visiting seminary students, her pastoral ministry deals with patients' spiritual needs "on whatever level they happen to be operating at the time," and Georgie J.'s eclectic, ad hoc spiritual needs have often operated on levels only God could fathom.

While in the chapel, Pascal also picks up a bottle of holy water requested yesterday by Paul B., a fifty-two-year-old ex-prisoner who is being released tomorrow after a three-month stay for treatment of non-Hodgkin's lymphoma. Although weak and emaciated, Paul has actually stabilized for the time being, thanks to intensive chemotherapy and radiation therapy. His fevers have abated, at least for now, and the tumors in his lungs and brain have actually shrunk. When Pascal drops off the holy water in his room late this afternoon, Paul is already packing his things for the trip home to Queens. He is ecstatic about tomorrow's discharge and all day long has intermittently been singing to himself the *Wizard of Oz* ditty "Ding Dong! The Witch Is Dead." Whether Paul realizes that his immunoblastic lymphoma is not even in remission, but only temporarily halted, is not clear, although he is definitely not demented. All he does know is that, after five years in prison and three months in St. Clare's, he is finally going home for Christmas with his wife and five children. For him, the wicked witch of AIDS is dead, if only for a few weeks, and he is savoring happiness with the gusto of a starving man enjoying his first good meal in ages. Today, Paul is so possessed with joy that he barely thanks Pascal for the holy water and returns to rearranging his things in his suitcase, merrily singing to himself.

Pascal also stops by the Spellman volunteers' office to request some extra magazines and games for the patients on the prison ward. The VCR and TV in the prison unit's lounge were stolen last weekend, leaving the inmates few diversions. It is still a mystery to everyone how those items could have been stolen from a secured, guarded unit.

Almost forgetting Israel T.'s request earlier this morning, Pascal asks Gary H., one of the Jesuit seminarians rotating through pastoral care this month, to pick up a get-well card for Israel T. to send to his sick son in Buffalo. Pascal briefly tells Gary about Israel's predicament and suggests he might want to talk with Israel when he delivers the card tomorrow.

"Just let Israel do all the talking. Don't feel you have to say anything, unless you really have something to say," Pascal hastily counsels Gary, adding that "Israel will find peace his own way, on his own terms."

Finally, postponed until tomorrow—or perhaps even the day after—is the task of replying to several inquiries recently received from pastoral care workers from across the country, who have written Pascal for guidance about their own HIV programs. And a couple of other minor administrative matters must also be put off until she has time for paperwork, because today she has one final visit to make, an important visit to Felipe A.

Felipe A. is a twenty-seven-year-old Puerto Rican man from the Bronx who was recently paroled from prison and who is dying of AIDS.

Paralyzed for several months from the waist down by a spinal-cord infection from a herpes virus, Felipe has suffered many of the complications of prolonged confinement to bed: deep and infected bedsores, recurrent pneumonias, intermittent blood infections from intravenous catheters, and contractures of arm and leg joints, which are now frozen in place and cannot be moved. Felipe is essentially skeletal, weighing only seventy-nine pounds; prison records documented his weight as two hundred five pounds only one year ago. Felipe has been totally blind from CMV retinitis for eight months, ever since a doctor in prison misdiagnosed his then blurred vision as due to "psychogenic hysteria."

Felipe's most recent problem is yet another pneumonia, and his breathing has become so labored that earlier today he was started on both oxygen and an intravenous morphine drip, the latter to ease his apparent breathlessness.

However, it is not possible to tell exactly how Felipe feels—whether he is in pain or is having difficulty breathing—because he has been in a coma ever since he was transferred to St. Clare's three weeks ago from a rural upstate hospital, where, four weeks earlier, he had been admitted from Attica prison. When he was given medical parole three weeks ago, the upstate hospital expeditiously transferred him downstate to the Spellman service. His parole meant the state was no longer picking up the tab. Evaluation of Felipe's coma at St. Clare's has been fruitless as to an exact cause, but most likely, it is due to CMV infection of his brain. Regardless of the cause, Felipe's brain is essentially gone, except for lower brain-stem functions that still maintain his basic vital signs.

Although his medical care at St. Clare's was thorough—the anti-CMV drugs ganciclovir and foscarnet have been used, but without any effect—Felipe's condition was likely hopeless by the time he reached the Spellman service. Indeed, he has really been reduced to a nursing problem, requiring total and constant care, with repositioning every few hours and frequent changing of his bedding, so his bedsores at least do not worsen. The medical staff's job has likewise been one of reacting to the myriad complications that ensue when a malnourished AIDS patient cannot move and is in a coma.

Felipe's mother abandoned him at an early age, and he was raised in a series of foster homes. It is not known if he has any brothers or sisters, and he has no known friends. Except for court and prison records, Felipe's life is essentially a blank, as if he never existed beyond the obligatory data generated by the bureaucracy. Fortunately, he was made DNR by a court-appointed health-care proxy while he was still a state prisoner. Whether that paperwork is still valid now that he is paroled is a matter no one wants to investigate, since doing so might jeopardize his DNR status. Felipe's condition has worsened considerably over the past twenty-four hours. His vital signs have gradually deteriorated—higher and higher fevers, more rapid respirations, and increasingly erratic heart rate.

Pascal seems to tiptoe almost reverently into Felipe's room. She has visited him many times over the past three weeks, even though they have never had a conversation. Indeed, it is doubtful he is even aware when anyone is there with him. Pascal first approaches Felipe's bedside, and seeing that he is asleep—or, more accurately, still comatose—she steps back and stands quietly in the dimly lit private room, her hands clasped in front of her.

Felipe's room is strangely peaceful this late mid-December afternoon. The tranquil bubbling of the oxygen through the mask on his face and the quiet humming of the intravenous-fluid machine somehow perfectly meld with Felipe's shallow, gurgling respirations. Outside the window, wet snow silently accumulates on the quiescent air conditioner, and far beyond the hospital's inner courtyard, several police sirens wail distantly in the concrete caverns of the big city surrounding St. Clare's. Only three crosstown blocks to the east, the late-afternoon performance of the *Radio City Christmas Spectacular* is letting out, and throngs of holiday shoppers are clogging the department stores and restaurants of Midtown. Except for the dwindling skylight diffusing through the grimy window, the only illumination in the room is coming from the partially opened bathroom. Felipe's bed is in the lit part of the room, and save for his shriveled-up head, his emaciated body is blanketed by white sheets, his wisp of a frame dissolving into their folds.

Pascal glances up at the only decoration in Felipe's room, a wooden crucifix one of the seminarians put on the wall next to the bed last week. It is a stark crucifix, not palatably sterile like the others in the hospital. The Christ on Felipe's cross is not gracefully muscular, with handsome Caucasian head turned downward in tranquil repose, with hands and feet antiseptically pierced by discreet, unobtrusive nails. Rather, Christ's emaciated, welted body is grotesquely twisted on the cross, the upward-turned face contorted in agony, and fractured hands and feet hideously splayed open by large, unsightly bolts. This somber crucifix reminds Pascal of similarly realistic crosses fashioned many, many centuries ago by fearful believers facing the Black Death.

After standing at bedside for a few seconds, Pascal, who is tired from all of her ministrations, turns and sits in a chair that is right behind her, in the shadows, facing Felipe's bed. Hands folded in her lap and legs crossed, she sits back and softly exhales a pensive sigh. Her eyes look down and are unfocused; she appears to be gazing into space, or perhaps into the many past years she has held similar vigils for similar patients. Half-entranced, half-meditative, Pascal seems intensely serene, intensely prayerful. It is as if her very soul could be read from her face during these important moments, when she is standing watch over Felipe A., like an eternal sentinel, a silent witness to the suffering of AIDS.

At peace in her thoughts for several minutes, Pascal is finally aroused from her reverie by the familiar sounds of Christmas carols wafting through the outside hallway. A local church choir, visiting the Spellman wards late this afternoon, is singing "It Came Upon a Midnight Clear." Pascal gets up from the chair and, in an almost involuntary gesture of both farewell and benediction, approaches Felipe's bed and, ever so slightly bending toward it, reverently touches the bedsheet with her outstretched right hand, bowing her head briefly. It is as if she were in the presence of something, or someone, sacred, holy. As she exits, she pauses at the door and looks back toward Felipe, again ever so briefly, then turns and leaves, her face suffused not with sadness, but with amazing grace.

Her rounds finished for today, Sister Pascal bundles up in her winter coat and quietly leaves St. Clare's for her solitary trip to her convent in the Bronx.

Throughout the ages, religions—especially Judaism and Christianity—have grappled mightily with the issue of faith versus good works as the pathway to personal salvation. On Pascal's daily rounds, the celebratory reverence she brings to her good works evokes the powerful assertion, "The Body of Christ has AIDS." This electrifying idea, breathtaking in its implications, raises timeless questions about God, humankind, and suffering—questions about the intersection of the eternal with the mortal, the infinite with the finite.

The Gospel writer idealized this divine presence in the suffering world as the Logos—the Word. Indeed, AIDS has clarified the duality of this divine presence in the world, a paradoxical duality that ultimately unites the caregiver with the care receiver, the living with the dying. The pastoral care of Sister Pascal—especially, the transcendent image of a dying Felipe A. and a beatific Pascal Conforti at his bedside—makes manifest the mystical duality of the Gospel's message that "the Word became flesh and dwells among us."

6

LAST WORDS

"Saturnino H." . . .
"Pedro V." . . .
"Maria M." . . .

Sometimes, a name called out from the altar is too familiar—uncomfortably familiar . . .

ON AN AIDS ward, sickness and death strike not just the patients and their loved ones. Indeed, few people realize that staff, too, can become victims, whether through burnout or physical illness. Such was the case of a physician assistant whom I worked with in the winter and spring of 1993.

Pedro, a new Spellman physician assistant, had been trying the best he could to adapt to the demanding responsibilities of working on an AIDS ward, before suddenly falling ill from pneumonia—a pneumonia so severe that he had to be hooked up to a life-support breathing machine. Not much was known about Pedro's personal life. His home was originally Peru, he had moved to America five years before, he said he had been married and had no children, he had recently graduated from a local PA college, and Spellman was his first job after PA school. A quiet, handsome man in his early thirties, Pedro largely kept to himself and did not seem to have many personal friends, although he was always congenial and accommodating. His shy personality reflected an engaging gentleness that, neither effeminate nor ingratiating, denoted an innocence of temperament that is not often seen in the chaos that is Spellman.

Pedro originally came to the hospital at the beginning of 1993. He never complained about his workload and never gossiped about coworkers or job-related politics at Spellman, which set him apart from other Spellman employees. Patients commented on his attentive ways, and the nursing staff appreciated his nondemanding demeanor. As the attending physician supervising Pedro on Unit 3A, I, too, liked his earnest, friendly manner with patients and staff. I felt that, despite his mediocre PA skills, he had the two major attributes needed to become a good PA—namely, he was educable, and he had no "attitude problem" toward work. Pedro would be no superstar, but with time he could become a solid player on the Spellman staff—that is, until unexpected illness suddenly ended his life and underscored uncomfortable issues of mortality for his colleagues.

Shocked, dumbfounded, incredulous, numbed—these were the reactions on Pedro's Spellman floor that sunny spring Monday, approximately five months after he had come to the job, as word quickly spread that he had been admitted over the weekend to a Brooklyn hospital and had been intubated for respiratory failure due to severe pneumonia. Rumors circulated that Pedro was comatose. According to reports, he was being treated with intravenous Bactrim and steroids, an indication that the presumptive diagnosis was PCP. There had been no indication, no warning, during the previous weeks that Pedro had been ill—no telltale cough or breathlessness, no specific complaints from Pedro that he was not well. A few of the nurses recalled in retrospect how he had recently remarked to them about tiring more easily, but such complaints have always been endemic to Spellman staff, given the ceaseless demands of the job.

As with the other AIDS units at St. Clare's, Pedro's floor had always had a close-knit sense of family among the staff, most of whom were on a first-name basis.

"How can he be gone?" one of the nurses asked me in disbelief that unsettling day. "He was just here last week," as if his absence was more significant than a patient's death, which was more routine.

Of course, she knew—at least intellectually—that someone can become seriously ill in an instant, let alone over a weekend. But the void of Pedro's absence that Monday in late May, and for many days thereafter, was a haunting reminder of our own fragile hold on good health.

The unfolding story about Pedro that May day was but one of several calamities that had befallen the Spellman service over the prior few months. Three weeks earlier, a well-liked clerk on another Spellman unit had died after a long bout with TB. Moreover, the Spellman doctors and PAs on the prison unit were trying to cope with the grisly death of one of the unit's nurses. The nurse, who had been out sick for weeks with an unknown malady—everyone suspected job-related TB, or even worse—was killed, either by bizarre accident or by suicide, when an Amtrak Metroliner ran over her in New Jersey. The mysterious circumstances of her death, plus the speculations about her illness, had kept her bereaved colleagues on edge. One of the Spellman PAs commented to me how these losses tempted him to stop making any "work friendships," since it seemed that only sadness resulted. Pedro's sickness, which seemed equally mysterious, made me reflect on the complex working relationship I had had with Pedro over the prior months. But my attempts at introspection were soon interrupted by yet more bad news.

For, on that same day, two other Spellman employees—a physician and an office administrator—had just come down with chicken pox, undoubtedly a consequence of their exposure to varicella, the chicken pox virus, which frequently afflicts Spellman's patients. Although usually not serious, chicken pox in adults can be uncomfortable and require up to several weeks of homebound convalescence, but in rare cases it can progress to potentially fatal varicella pneumonia.

My stomach, in fact, churned with anxiety, a feeling brought on by Pedro's intubation but exacerbated by the outbreak of chicken pox. Although it was still unclear whether Pedro's illness was acquired at St. Clare's, the possibility of job-related tuberculosis—even varicella pneumonia—was on everyone's mind, especially with the recent deaths of the Spellman clerk and nurse. To deny that an AIDS ward could be a dangerous place to work would be foolish at best.

As that stressful, almost surreal day plodded along—as worried colleagues periodically asked me about any more news of Pedro—I tried hard to recall my interactions with Pedro over the previous weeks, to see if I had failed to pick up any telltale clues. I tried to remember our last words together, as well as any signs of impending illness that the pace of work might have caused me to overlook. I wondered what had happened over the previous weekend to my enigmatic physician assistant. I tried to imagine what he must look like in intensive care, dependent upon life-support systems. Was he conscious, did his family visit, did he have thoughts about his patients and colleagues at St. Clare's? Had his prior five months' work with sick and dying AIDS patients provided him with comfort or increased his fear in the face of his own present suffering?

When serious illness hits one of the Spellman staff, the sobering realities of death and disease become uncomfortably personal, like stark, bony fingers to the heart. Numbed by the daily challenges of caring for dying AIDS patients, even the most dedicated staff members are caught off guard by their own or a colleague's illness, an eventuality best epitomized by George Bernard Shaw's dictum that there is nothing more pitiable than a doctor who is sick. For health-care workers, especially at an inner-city hospital such as Spellman, the all-too-human denial of death's ubiquitousness often degenerates into bravura, even arrogance, as some try to insulate themselves from the fearsome realities surrounding such issues on the job. Moreover, when the job has not insignificant risks of acquired diseases—sometimes fatal diseases—a whole new dimension of fear is added to the other stresses of caring for diseased and dying patients.

The on-the-job health risks of Spellman's AIDS wards vary in seriousness and frequency and are much different from what the general public might think.

Contrary to popular belief, accidental infection with HIV is *not* one of the major occupational risks of working with AIDS patients. Acquired in health-care settings only by needle-stick injuries, HIV is transmitted by needle-stick only one-third of one percent of the time. In spite of due vigilance, needle-stick injuries occur regularly on Spellman wards, and whenever a Spellman worker has such an injury, there arises the primal fear, usually unspoken, that the split-second needle-stick may irrevocably transform the caregiver into one of the eventual care receivers—that the invisible line between being HIV-negative and being HIV-positive has been transgressed. Because it can take three months, or even more, of periodic HIV testing before being certain of no transmission of HIV, a hospital staff member has many months—and often many sleepless nights—to ponder the imponderable what-if questions: "What if I'm HIV-positive, what if I get sick, what if I develop complications of AIDS, what if I become sick like the patients I care for at work each day—what if I become blind or demented or emaciated or incapacitated by diarrhea or disfigured by KS—*what if I die from this needle-stick injury?*" More aware of the physical devastation that the illness can bring, the caregiver may have fears that are both pronounced and acute. Fortunately, however, there have been no known instances of HIV transmission to a Spellman worker from a needle-stick, but the psychological stresses of such events are immense.

But not all health risks on Spellman are as awesome as HIV transmission from needle-stick injuries. Probably the most common job-related infections on the Spellman service—scabies and body lice—are more of a psychological setback than anything life-threatening. Because so many

Spellman patients sadly lack appropriate body hygiene—sleeping in the subways or in cardboard boxes on the sidewalks is not conducive to pristine cleanliness—body lice and scabies are often admitted to the hospital bed along with the patient, and the vexing little critters, who respect no socioeconomic barriers, love to jump unfettered from patient to nurse or doctor, much to these caregivers' annoyance and embarrassment. Anxious to the point of being neurotic about taking these skin infestations home, overly zealous staff members on various Spellman floors have periodically whipped themselves into scabies frenzies—group hysteria much like a religious revival meeting—in which the slightest itch or scratch by patient or staff member was deemed indicative of skin infestation that required immediate purging with Kwell, an antiscabies lotion, accompanied by an almost ritualistic laundering of all clothes and bed linen. During these "epidemics," it would not be unusual for there to be multiple purifications of the same patients and staff members. Although such skin infections are not pleasant to get—let alone take home to the family—they are completely curable. Nonetheless, nothing is more likely to excite, and incite, the staff than a good old-fashioned scabies outbreak, despite the much more serious infections floating around the Spellman floors.

However, two major HIV-related infections do pose serious risks to caregivers on AIDS wards such as Spellman's: hepatitis and tuberculosis, both of which have afflicted several staff members episodically over the years.

Hepatitis B and C are viruses that can be transmitted via needle-stick injuries much more easily and more frequently than HIV. Commonly found in drug users, both viruses can cause acute hepatitis of moderate to severe intensity, oftentimes laying up their victims for weeks to months before gradually resolving. However, the major side effect of these viruses is chronic liver damage that can cause not only debilitating fatigue but also chronic cirrhosis and even liver cancer many years later. Several Spellman staff, including one physician, most likely acquired hepatitis C from otherwise benign needle-stick injuries, with resultant need for periodic blood tests and treatment with injections of interferon, which itself causes deleterious side effects. Yet the lack of social stigma associated with these potentially serious liver viruses has often—quite inappropriately—lulled the Spellman staff into regarding them as minor consequences of needle-stick injuries.

The one highly contagious disease on the Spellman wards that does cause the staff to pause and tremble is tuberculosis, probably the most serious health threat to employees of all kinds at St. Clare's. Merely walking into the hospital as a visitor, let alone staying there for any length of time as an employee or patient, significantly increases the risk of infection with this highly airborne, potentially fatal disease. Moreover, AIDS predisposes a person exposed to TB to develop active, often severe disease, which accounts for the simultaneous epidemics of HIV and TB. A frightening twist to this story has been the emergence of multidrug-resistant strains of TB—MDR-TB—which are impervious to many of the previously effective, first-line anti-TB drugs. Since the early 1990s, St. Clare's has been in the vortex of the urban MDR-TB epidemic, which only a few years ago threatened to depopulate the country's major cities. Indeed, on the basis of epidemiologic studies, the hospital even has the dubious distinction of having a particular subtype of this TB "Andromeda strain" named after it.

Given the highly contagious nature of this disease and the large number of HIV-infected patients who have it, it is not surprising that so many St. Clare's employees show evidence of infection with TB, as indicated by their having a positive TB skin test. In 1991 and 1992, TB struck the Spellman medical staff in a significant way: one physician died from TB meningitis—a devastating infection of the spinal fluid and brain—and two other physicians and one PA came down with active lung TB, one case of which was MDR-TB. Fortunately, these latter three casualties recovered, but only after many months of bed rest and even longer periods of taking unpleasant medications. At the beginning of this mini-epidemic, the hospital administration foolishly contended that these TB cases among the staff were not job-related—"they could have picked it up on the subway," speciously argued one of the hospital administrators. As grim testimony to its ubiquity at St. Clare's, TB has struck not only the clinical staff but also employees not directly involved in patient care, including a clerk in the admitting office, a secretary in the administration office, and an elderly nun in the social work department. Sometimes the patients themselves have ironically acquired this pernicious microbe simply by being at St. Clare's.

During St. Clare's TB rounds—the weekly conference with visiting infectious-disease experts—staff members often discuss those patients who have had to endure week after week of respiratory isolation. Never articulated, however, is the silent fear that someday one of the doctors or PAs at those rounds will come down with TB and end up in respiratory isolation, stripped of autonomy, shut off from the world of his or her healthy colleagues. The Spellman worker's vulnerability to TB always permeates the wards of St. Clare's, an ever-present specter threatening to transform the unsuspecting doctor or PA into just another TB patient. This unspoken vulnerability is but another reality that lessens the chasm between caregiver and patient at St. Clare's.

After news about Pedro's illness first broke, only a smattering of information followed, and the paucity of facts fueled speculation that he had been striken with TB or varicella. However, worries about job-related diseases aside, the anxiety among the staff really went much deeper, to primal issues of mortality, especially mortality of the caregiver.

Gradually, information about Pedro's condition surfaced. He apparently remained intubated and heavily sedated, totally dependent on the breathing machine. One of the Spellman administrators revealed how Pedro's preemployment physical exam four months earlier had shown thrush—often a harbinger of immune system dysfunction—as well as prior hepatitis C infection. "I told him he might be HIV-positive and advised him to get HIV-tested," the administrator clucked, "but he refused." This startling revelation—that Pedro knew at the time he started working on Spellman that he might be HIV infected—added greater poignancy to his plight and heightened my worry about him.

After work on Friday that first week, I trekked out to the Brooklyn hospital to visit Pedro, to see for myself what I still found difficult to believe. Sure enough, there he was, lying motionless and unconscious in his ICU bed, his chest gently heaving up and down with each rhythmic sigh of the breathing machine. The standard panoply of tubes, monitors, and IV lines covered his body—the veritable "full-court press" of high-tech critical-care medicine. Although bloated from the IV fluids, he was still quite recognizable through the puffiness as Pedro, my PA, who only a week earlier had been busy on his AIDS ward, drawing blood, starting IVs, chasing down lab and X-ray reports, cajoling patients into taking their medications, translating Spanish for me, and comforting patients and their families. Although I knew it was Pedro in that ICU bed, it still seemed unreal to me, as if this could not really be happening. I was infinitely more affected by the sight of him than the reality of any of my patients on 3A, as if he were a direct link to me—and therefore so suddenly did I become much more vulnerable, more compromised.

The ICU staff were pleasant and even grateful for my visit. They said Pedro's brother had been his only visitor all week, coming early every morning for ten minutes or so. Pedro's nurse reported that he had remained unconscious since the previous weekend, when he had developed respiratory failure requiring intubation. Before leaving, I left a short note at his bedside, addressed to his family, in which I said I had visited Pedro and wanted them to call me if there was anything I could do.

I left Pedro's hospital bed that evening still in disbelief, as well as with wistful regret that I had not known him better when we were working together on our AIDS ward. Once again, as it had done several times over the previous days, my mind raced through the events of the prior weeks and months. As with so many other people in my busy life, I had largely taken Pedro for granted, not making the effort to know him better.

Perhaps it was that he had only worked at Spellman for five months, but Pedro had been a difficult person to get close to. He was reserved, almost to the point of being mysterious, but he did have a special sense of humor, a delightful blend of deadpan and droll understatement. During our busy days, we would regularly commiserate about the day-in, day-out absurdities of trying to provide rational medical care in this veritable nuthouse. But, unlike most Spellman staff—many of whom, consciously or unconsciously, would come to Spellman to work through their own psychotherapeutic issues—Pedro did not share much of his private life with anyone. Despite his reticence, Pedro had struck me as generous and bighearted, albeit in a quiet way. He always offered to share his snacks with me and the other ward staff, and he would often buy treats for his patients. When I once casually told him how a close friend of mine was moving to New York and would be looking for a job as a waiter, he offered—and thereafter repeatedly reminded me of the offer—to put my friend in touch with acquaintances from his own days as a waiter while in PA school. But despite his warm manner, he was always oddly vague and evasive whenever I would ask about his outside interests, his wife, or any other family life in New York. Pedro was ultimately an enigma, a man of few personal revelations.

During that early-evening taxi trip home from Pedro's ICU bed, I once again—for what seemed the hundredth time that week—wondered if I should have said or done something more for Pedro in the weeks prior to his illness, weeks that had not been easy for either of us. Because of a family death, the other, more experienced PA on the floor had been absent for the three weeks preceding Pedro's illness, and there was no one to fill in for him. Thus, Pedro and I had to take up the slack, and because he was still relatively inexperienced, the major responsibility for managing the patients was mine, while Pedro did the scut work—drawing blood, starting IVs, and all of the other mindless chores of patient care. Both of us felt the stress as the weeks droned on. I felt emotionally enervated, much like a zombie, from seeing all the patients and writing all the progress notes every day, and Pedro grew increasingly harried from the endless tasks I continually requested of him. By the second week of this ordeal, Pedro's performance began a gradual decline, which was at first just annoying and later became worrisome.

Even before he had to assume the work of two PAs, Pedro had never been very efficient. He had always seemed distracted, a little too prone to forget things I would ask him to do. When the other PA was there with him, this absentmindedness had been manageable, but when it was just the two of us, his seemingly lackadaisical, forgetful manner irritated me, sometimes to the point of my scolding him. He never complained or made excuses about his lapses, always apologizing and promising to try harder. But the mistakes and oversights continued with increasing frequency. He would ask the nurses for the dosages of medications he was ordering, information he should have known at that point in his career, and they sympathetically assumed he was overworked and helped him out as best they could. It seemed that the harder Pedro tried to avoid mistakes, the more error-prone he had become, and I often had to remind him three or four times to check this or that test result, to do this or that chore. Yet, despite the work and despite my complaints, he would still maintain his amiable nature, never telling anyone that he felt ill or was having personal problems. I had never yelled at him or berated him, but as his performance became sloppier and less efficient, I made it clear to him that he was not up to par, that he was simply not good enough.

Now that Pedro lay in extremis, I regretted that I had not taken him aside and asked him if anything was wrong, if there was anything I should have known about his health or personal life. As I rode back to Manhattan from Pedro's ICU bed in Brooklyn, I self-critically replayed the previous weeks in my mind's eye, and I had to agree with the reassurances the ward nurses had been giving me over the past week of soul-searching—namely, that Pedro really had not had any overt, obvious signs of impending sickness, that except for appearing understandably tired and hassled, he had not given out any signals, hints, or suggestions that he was about to crash into a life-threatening illness. While I regretted I had made Pedro feel inadequate, not up to my high standards, another side of my conscience—the self-protective side—reminded me that, during those difficult weeks, I had been running an AIDS ward, not a psychosocial support group for stressed-out PAs. I also knew that if Pedro had come to me for help, I would have done all I could to assist him, to encourage him to rest and call in sick. But his condition still vexed my conscience. Perhaps, just perhaps, if I had not been so gruff, so overwhelmed by work responsibilities, so focused on my own misery from the extra work on the ward—perhaps if I had been able to pause and approach Pedro as a friend rather than a supervising physician—then maybe, just maybe, he might have confided in me.

During that taxi ride back from his bedside, I reflected on some of the terminal patients Pedro and I had cared for, wondering how he must have felt seeing them suffer and die, especially since he *must* have sensed that he, like his patients, was very sick. There was Maria M., a forty-year-old, bedridden, comatose woman who was profoundly emaciated and had been lingering on for months and months, despite not eating or drinking much of anything. Everyone involved in her care was absolutely amazed how she clung to life for so long, since by all medical standards she should have been dead months ago. I now winced recalling how I had privately joked several times with Pedro about how Maria was "the woman who wouldn't die." And then there was Saturnino H., a young man from Colombia who had overwhelming cryptococcal meningitis and had, shortly before death, lapsed into a coma. Because his family had adamantly insisted he be resuscitated, the macabre ritual of trying to revive a skeleton eventually had to be played out to satisfy New York State law, and Pedro was there with me at bedside as the two of us cynically pantomimed the slow-code routine of "resuscitating" Saturnino's bag of bones, which had the good sense to remain dead. Now that Pedro was every bit as sick as some of his previous patients, I wondered about the past emotional pain he must have had when he would see his patients suffer and die, sensing—as he *must* have—that he was soon to follow in their footsteps.

Yet I also realized that such musings raised subtle, yet profound existential issues. After all, in the ultimate context, as the two of us cared for the dying on our AIDS ward, Pedro's situation and mine had been the same: *both of us were going to die, and the only difference—philosophically, a trivial difference—was that Pedro should have sensed he would be dying soon, whereas I had no idea whether my death would be sooner or later.* It suddenly occurred to me on my trip from Brooklyn that I could still die before Pedro did and *would probably be less prepared than he would be.* Startled at that sobering thought, I glanced at the accelerating speedometer of the rattling taxi speeding me back to Manhattan and nervously barked out to the driver to slow down.

By the time I arrived home that Friday evening, I had tried to work through my self-doubts and self-criticisms. There was no way to know whether Pedro would ever have turned to me for help, and I reflected with satisfaction on how one of the many blessings of growing older was learning not to be too hard on myself in cases such as this. But second-guessing myself aside, Pedro's story highlighted a crucial verity of which I always needed to remind myself. Namely, as trite as it might sound, I never know when my last words to someone might indeed be my last words. Mulling over this simple insight, I tried to remember my last words to Pedro one week earlier, at the end of a hectic day that was ending yet another interminably exhausting week on our AIDS ward. All I could recall was how relieved I was that the week was over and how besieged my head felt from the onslaught of the prior five days. Try as I could, I still had only a vague recollection of bidding Pedro good-bye that Friday. He was standing at the nursing station, finishing up some paperwork, and when he assured me everything was okay, I told him to finish up and go home. As best I could recall, he looked the same as he had for the prior week—slightly harried, but still standing after a week of interminable work.

Over the ensuing several weeks, Pedro gradually slipped from the forefront of everyone's consciousness, including mine. I still thought about him daily, and nurses, physician colleagues, and other PAs still asked me for any news, which remained bleak: he remained unconscious and dependent on the breathing machine. The Spellman assistant medical director was in contact with Pedro's doctor, who revealed that tests had diagnosed AIDS and *Pneumocystis carinii* pneumonia, thereby confirming everyone's fears. The ease with which colleagues discussed and communicated Pedro's medical condition bothered me, the final details of his heretofore private life being open to anyone who inquired. Other fragmentary information trickled in from the few nurses and social workers who also visited Pedro in ICU. Apparently he had been separated from his wife, and he had become estranged from his family in Peru for several years because of undisclosed disagreements about his personal life. There was also vague information that he had been sick in the past, possibly from meningitis. These bits of Pedro's past raised tantalizing questions, which ultimately remained unanswered.

Surprisingly, Pedro's brother telephoned me at home exactly two weeks after I had visited Pedro in ICU. Thanking me for visiting his brother, he reported that the doctors were increasingly pessimistic about Pedro's chances.

He confirmed that he was Pedro's only family member in America but said their sister was on her way from South America to represent the rest of the family. I heard a genuine affection for Pedro in his voice, and his gratitude for the prayers and support of the Spellman staff seemed heartfelt. He also sounded a little forlorn and confused, as if he did not really understand the enormity of what was happening to his younger brother. Sensing that Pedro's brother had called just to talk with someone about his brother's illness, I went on to elaborate on how sensitive and helpful Pedro had been on 3A, and I urged him to keep in touch with me. After the call, I actually felt relieved that Pedro had had someone like his brother in New York, and I sympathized with the difficulty his brother apparently had in dealing with this crisis, a catastrophe neither of them had anticipated when they arrived together in New York many years ago.

By the third week of Pedro's vigil, there was no new information, and the shock of Pedro's illness had run its course. With the recovery and return to work of the two Spellman staff with chicken pox, the routine of the AIDS wards was again undisturbed by unpleasant reminders of how even the caregivers are not immortal. That Pedro's disease was probably not acquired on the job also dissipated everyone's anxiety. He was thus relegated to being just another patient, not a colleague whose suffering might someday be our own. Indeed, whenever new information about Pedro would surface at work, the reaction was no longer one of sympathy but rather an unbecoming smugness.

"Pedro must have been in denial about his illness, otherwise he would have gotten HIV-tested and sought medical care," one of the 3A nurses chimed in when news about his prior episode of meningitis came up.

"In denial"—the critical label the healthy glibly give the sick whenever the sick do not behave exactly the way the healthy think they should. Nothing is more arrogant than the condescending pop psychology the healthy spout off about the so-called denial of the sick, when, in fact, most of the time it is the smug, self-assured healthy who are in denial—in denial about their own fragile mortality. We all like to believe that we will live in perfect health until the age of ninety-six and then die tranquilly in our sleep, and thus we have erected a culture that so diminishes the wisdom of the elderly and eschews the sick and dying. Eventually, however, everyone—often sooner than we believe is possible—will cross over from complacent good health to worried sickness, sickness that is often endured the same way good health was experienced—in denial.

Pedro's death, from what I was told, was quiet and uneventful, three weeks and one day after he entered the hospital. His brother called me at home late that night, just before midnight, to report the news. He said that he had instructed the doctors to withdraw medical treatment earlier in the day, since they had advised him there was no hope for recovery. Pedro's sister had arrived and was at his bedside when he died. There was the same bewildered disbelief in his voice that I had sensed on our prior conversation—an unspoken feeling that this tragedy could not really be happening. Expressing my regrets, I gave the same reassurances I had given countless other bereaved family members over the years: namely, that Pedro did not suffer when he died, that death was undoubtedly painless. His brother thanked me and said Pedro would be cremated, his ashes to be sent later to his home in Peru.

I reacted to the news without much emotion, but had an odd feeling of inconclusiveness—a sadness about my not getting to know Pedro well before his unexpected illness. If his sickness had been more drawn out—if he had not been unconscious and hooked up to a breathing machine—then maybe I would have had one more chance to talk with him, to extend a hand of friendship. Everything had happened far too fast, without warning, and both his innate reticence and the hectic pace of the ward right up to his illness left everything a blur in my mind. Having known from almost the outset that Pedro would not recover, I had emotionally said good-bye to him weeks before, although I could still not remember the exact details of my last words with him almost four weeks earlier.

A little less than two months after Pedro's death, Sister Pascal and two of the Spellman social workers arranged a special memorial service for him in the St. Clare's chapel, at 4:30 P.M. on a Thursday. Flyers announcing the service were posted throughout the hospital and were put in the mailboxes of the Spellman staff. However, other than Father Jack and Pedro's brother and sister, only eight people showed up: Sister Pascal, three social workers, a nurse from Pedro's ward, an elderly lab technician, my best friend, and myself. No other physician assistant, physician, or anyone from Spellman administration attended, everyone apparently too busy, too forgetful, or perhaps too embarrassed to remember a colleague who had died two months earlier from AIDS. I opened the service by noting how Pedro's suffering and death reminded all of us of our fragility—how, despite our white coats and professional titles, our time will come someday, just as it had for Pedro and many of our patients. I closed my remarks with fond remembrance of Pedro's gentle ways, describing for his family his quiet caring for the disadvantaged with AIDS. Father Jack then officiated at a mass in Pedro's memory. His homily emphasized the special joy Pedro gave his patients, but like my remarks, it did not mention that Pedro died of the same disease afflicting his patients. The twenty-five-minute service was simple, even by Spellman standards, but his brother and sister genuinely appreciated the kind words about their younger brother. The slight attendance probably bothered me more than it did Pedro's brother and sister; they were just grateful that someone thought enough of their brother to arrange such a service.

Walking home from work that evening, I reflected on how it unfortunately took someone's unexpected death to remind me how special that person was when still alive. When alive and well, Pedro did not seem to be a memorable person, but his death deeply touched me. And once again I tried hard to remember my last words with my now largely forgotten physician assistant . . . last words that, unknown to both of us, were to be . . . my last.

7

ROOM 305: A SPELLMAN SUITE

"Norman W." . . .
"Enrique P." . . .
"Elvis C." . . .
"Angel T." . . .
"Mabel C." . . .
"Judith A." . . .
"Jay M." . . .

So much crap, just to die.

—ENRIQUE P.

It's only *AIDS, honey, not something* really *serious.*

—ANGEL T.

OVER A THREE-month period in 1993, seven different patients successively occupied room 305 on my AIDS ward. At the end of these three months—by the next quarterly memorial service—six of these seven people were already listed in the Book of Remembrance, and the seventh person's entry into the memorial record of the dead was pending. Not remarkably, at least for the Spellman service, these seven people reflected the vast diversity of Spellman patients with regard to age, gender, socioeconomic status, and sexual orientation. There were a closeted homosexual Hispanic male, two ex-prisoners, a partial transsexual, an elderly woman, a young female crack addict, and an affluent gay white male—the spectrum of AIDS in New York City.

Just as these seven patients were, for Spellman at least, typically atypical in their demographics, so was the hospital room in which most of them spent their last waking moments of life. Room 305 is one of the more spacious rooms on 3A, measuring all of eleven by fourteen feet. Located close by the unit's nursing station, it is one of the few rooms that faces out onto the street and not onto a shadowy courtyard. Nonetheless, the street view from 305 might be disquieting for some—a somber-appearing funeral home sits directly across West Fifty-first Street. Since most of 3A's rooms are without any bath or shower, the greatest amenity of room 305 is the full bath and shower, which, however, must be shared with neighboring room 306 via connecting doors. Moreover, this shower must also be used by other less fortunate patients on the unit, and each morning a procession of 3A denizens files in and out of 305's shower, sometimes to the annoyance of 305's occupant. The separate sink and commode in 305 is crammed into a three-by-four-foot "bathroom," which has chronically suffered from a leaky toilet and a crumbling ceiling, which, mercifully, has not injured anyone. Except for these amenities and eccentricities, room 305 is otherwise quite typical of other rooms on the unit. The grime and dust coating everything within it is neither greater nor less than that of other rooms on 3A.

Yet beyond the demographics of gender, ethnicity, and income was the spectrum of emotional responses to AIDS that these seven "typical" patients displayed over the total twelve weeks' time they spent in room 305. Indeed, "So much crap" and "It's *only* AIDS"—these two radically divergent views about living and dying with AIDS were voiced within a few weeks of each other, by two profoundly different patients who expired in room 305. Enrique P. was an ex-inmate slowly fading away from TB, who uttered the above words with a disgust, a weary despair, about his lot. Verily, "So much crap, just to die" could also be a viewpoint on the entire burdensome affair of life itself, especially for a joyless AIDS patient such as Enrique. Angel T.'s outlook, on the other hand, denoted an approach to AIDS—and life—that eschewed rancor and self-pity and instead exuded grace and humor on the arduous journey to the grave. Dying a few weeks after Enrique, Angel was a partial transsexual who gloried in the moment, especially when that moment was defined in glittering drag costumes. Whereas for Enrique life was an undeniable ordeal, for Angel it was—sometimes literally—a veritable cabaret, in spite of AIDS. Yet even Enrique's ignoble, largely anonymous death revealed a previously hidden dimension to the man that impressed me deeply.

The twelve weeks these seven patients lived, and died, in room 305 encapsulated the universal Spellman experience, a rarefied microcosm of space and time. Definitely not Neil Simon's *Plaza Suite,* room 305 was more than just a procession of individual patient stories. Rather, this "Spellman Suite" witnessed over only a three months' span remarkable human experiences seldom seen elsewhere.

• • •

NORMAN W.

Visiting Norman W. on my daily hospital rounds was always a singular pleasure, an extraordinary lesson on how a person might bravely endure the countless ravages brought on by AIDS.

Norman was forty-one years old when he was assigned to room 305 for what was to be his final admission to the Spellman service. Although Hispanic, Norman had a Jewish last name. He said his grandmother had emigrated from Germany to Puerto Rico in the 1920s, but he knew nothing more about her background. Having worked most of his adult life in a coffee shop in Hell's Kitchen, he never married. Norman, in fact, had only his older sister, who looked after him when AIDS had weakened him to the point he had to stop working and go on disability. Norman's risk factor for HIV was never really discussed, but he had no history of prior IV drug use or blood transfusions. Most likely he was gay, but like many other Hispanic homosexuals, he probably felt forced to stay in the closet because of the homophobia so extensive in large parts of the Hispanic community. Norman once confided to me that he had no close friends, at least who were living, and his sister confirmed that all of his best friends had already died from the same plague that was now visiting Norman. Thus, except for his spinster sister, Norman was alone.

Admitted initially for bacterial pneumonia, Norman suffered one complication of AIDS after another. No sooner did the pneumonia resolve than a staph blood infection developed from a contaminated intravenous needle. Just as the blood infection was responding to antibiotic therapy, he became afflicted with profuse and watery diarrhea, probably as a consequence of the previous antibiotic therapy for the pneumonia and the blood infection. The bowel infection was then followed by a severe flare-up of painful mouth ulcerations that made eating torture. A second bout of a superinfecting bacterial pneumonia then promptly ensued, requiring more powerful—and more toxic—antibiotic treatment.

As one problem followed another, Norman became progressively weaker and more malnourished. He soon was able only to walk from bed to bathroom and back again. The recurrent pneumonias had seriously damaged his lungs and limited his exercise tolerance to the point that he often required bedside oxygen. The mouth sores extended to his lips and nostrils, which became encrusted with painful ulcers. Although the diarrhea could largely be controlled with medication, it was still incapacitating enough to cause occasional fecal incontinence, as well as more continual discomfort from bleeding hemorrhoids. The final indignity, the last insult for Norman, was a case of disseminated herpes zoster—shingles—which caused a painful rash of small sores scattered over his entire body. As these infected sores opened up and became tender skin ulcers, I was reminded of the Old Testament description of Job, "smote with sore boils from the sole of his foot unto his crown."

Despite one misery after another heaped upon him, Norman, like Job, did not "curse God and die." Instead of focusing on his suffering, Norman dealt with his misfortunes by turning and reaching outward—in this way deflecting his pain by projecting his concerns about others. Thus, he was always more worried about trying to assist his nurses and housekeepers with their daily chores, such as making his bed, cleaning the toilet bowl, or busing his meal trays out to the service elevator. He shared gladly with his neighbors on the ward what few things he had—his cassette tapes and tape player, his few magazines, his telephone, even his food.

Even more remarkable, he always made it a point to inquire after the health and well-being of *everyone* who came into room 305: not only the doctors and nurses, but also the housekeepers and maintenance workers, whom he cheerfully asked how their day was going, or how their previous weekend or holiday had been. And somehow Norman's regular inquiries always conveyed a genuine concern, a real warmth.

Almost always my daily patient rounds on Norman would have a remarkably poignant ending—an ending that was as predictable as it was gratifying. I would first go through my usual routine of asking Norman about his symptoms, then do a quick physical exam, and finally give him an update on his condition. After being sure he had no questions about his condition, I would then turn to leave, only to have him stop me in my tracks with his urgent-sounding, heartfelt question, "Yes, but, Doctor, how are *you* doing today?" Norman always projected a profound interest in me and my problems, as if my petty concerns were as important as the procession of afflictions pulling him closer and closer to the grave. Norman would never let me end my visit with him until I had assured him that I was fine and that my day was going okay.

Going into 305 to see Norman was like visiting an old friend. His outstretched hand and unfailing "Hello, Doctor" always accompanied my entry, until eventually the hand was too emaciated to reach out, and then, even later, the voice became too frail to speak, and the face too weak even to smile. But right up to the end, his eyes lost none of their humanity and gleam.

In the days before his death, Norman was barely conscious, but even then, after the medical business of my daily visit with him was finished, I would tarry at his bedside for a moment to stroke his head, to hold his hand, and to reassure him that I was just fine, thank you.

• • •

ENRIQUE P.

Enrique P. was transferred to room 305 from the Spellman prison ward the same day that Norman W. died. At the time, it seemed like just another routine case of an ex-prisoner who had just been paroled and would die soon from end-stage AIDS. My counterpart on the prison unit had called me immediately prior to the transfer, to fill me in on my new patient's case.

"Enrique's pretty far gone," my Spellman colleague concluded. "He's dwindled to the point he doesn't say much of anything anymore."

Receiving this bleak report, I again thought to myself how much I would miss Norman W.'s smile and enthusiastic "Hello, Doctor," which, according to my colleague, was to be replaced by Enrique P.'s unpleasant grunts and groans.

Instead, Enrique showed me how unconditional love can be.

Imprisoned in an upstate maximum-security facility for almost ten years, Enrique had been admitted five weeks earlier to the Spellman prison unit for fever and productive cough. Initial evaluation showed he had pulmonary tuberculosis, a bad diagnosis to have. HIV had long ago decimated his immune system, and the New York State prison system had festered for years with the MDR—multidrug resistant—strain of TB. And although there were other, lesser-known drugs that might kill these new MDR strains, these second-line drugs were often less effective and more likely to have toxic side effects. In Enrique's case, it was not yet certain whether his TB was the MDR strain—it would take another few months for the final cultures to come back—but because he had been a state prisoner, his Spellman doctors on the prison unit had to assume the worst and treated him for MDR-TB with a regimen of six medicines, informally nicknamed over the years "the St. Clare's Six." Indeed, St. Clare's had become a frontline hospital in New York's battle against MDR-TB, and weekly TB rounds with visiting TB experts carefully reviewed the progress and treatment plan of *every* TB patient on the Spellman service. Thus, Enrique P. was actually in one of the country's best places for TB treatment, with the most up-to-date therapy at his doctors' disposal. Nonetheless, when these MDR strains of TB infect someone with AIDS, the consequences can be catastrophic. Not only can the second-line medicines be more toxic, but the immune suppression from AIDS retards the patient's response to therapy. The most potent antibiotics in the world are ultimately useless unless there is an immune system to back them up.

By the time he was transferred to my AIDS unit, Enrique not only had no immune system—his T-cell count was literally zero—but he also had very little body left as well: he weighed only eighty-six pounds. With straggly, long brown hair and matted, dirty beard, Enrique looked twice his age of thirty-nine and was far too weak to care for himself, being thus confined to bed, where he was incontinent and had developed early bedsores on his buttocks and lower sacral area. Enrique's horrendously poor personal hygiene reflected the toll of long-standing self-neglect, relentless HIV disease, and overtaxed ward nurses, who only had time to clean up his feces-stained bedsores, but not his unmanicured, feces-stained fingernails or his food-encrusted, rotting teeth or his snot-filled nose and mustache. Enrique P. was not a pretty sight to behold or to smell.

I never really got a chance to know Enrique. Both HIV dementia and malnutrition had severely addled him to the point he knew only his name, and then only during those brief times he was awake. His waxing and waning alertness inexorably waned as his hospitalization wore on. The only time I ever heard Enrique speak anything at all was once on rounds, when I was struggling to prop him up in bed to listen to his lungs. Apparently weary from the interminable pokings and proddings of his then six-week hospital stay, he arose briefly from his chronic stupor to register his disgust with it all.

"So much crap, just to die." His voice, though raspy and feeble, bristled with exasperation and impatience, a terminal exhaustion that denoted singular insight into his condition. Thereafter, Enrique never again spoke to me, retreating back into the fog of his dreams.

Even when fed by a nurse's aide, he ate little and would often be unable, or unwilling, to take his TB medicines. Sadly, a good day for Enrique was one when he would be awake enough to pull out his IV line or his urinary bladder catheter. A bad day was one spent essentially in a coma, with incontinent bowels, gasping respirations, and cold, clammy skin. Fortunately, early on, when he was still halfway alert, his doctor on the prison ward had been able to convince him to sign a DNR, thus obviating the worry about trying to resuscitate him. By the time he reached 305, I could do absolutely nothing either to help or to hurt him. Although I "cared" about him in an abstract, almost generic sense, his already moribund condition on transfer prevented any emotional bond from being formed between us. To me, he was just another routine case of an emaciated, severely demented ex-prisoner with terminal AIDS and TB. Enrique was on what I termed automatic pilot.

Enrique's only family was his wife, Carol, whom I would periodically update about his condition. I never met her in person before Enrique's death, but over the telephone she sounded like a caring person. I remarked to Abby A., Enrique's social worker—and here I felt a little guilty in sounding elitist—that Carol P. sounded "normal" and "middle class," not at all like the wife of an of inmate of ten years such as Enrique. From her questions over the telephone, she seemed to understand both AIDS and the hopelessness of her husband's condition, and she would call me every few days for condition reports, since her job's daytime hours limited her hospital visits to the evening, after I would leave. Enrique's evening nurses reported that his wife would visit almost nightly, often helping the staff with his bathing and other personal care. Perhaps because of the uneventful aspect of Enrique's downward spiral, these brief telephone conversations with Carol were not memorable and likewise seemed routine. I supplemented my bleak appraisals with routine assurances that we would try to keep Enrique comfortable. Everything about Enrique's case seemed straightforward and boring—up until his death, that is.

Enrique P. died a little more than two weeks after his transfer from the prison unit, and three days after my last phone conversation with Carol. On my patient rounds the morning of his death, he was already in the agonal phases of the death rattle, with gasping, gurgling breaths of rapidly diminishing volume and rate. Normally I would have made a hurried effort to contact the family about such a drastic deterioration, so that the loved ones could brace themselves for the bad news or perhaps even come in before the last gasp. But Enrique's "automatic pilot" had nose-dived so rapidly that by the time his wife would have frantically rushed in, he would most certainly have already died. Better to wait a few minutes for Enrique to complete his death rattle, I reasoned, and then deliver the blow to his wife once and for all. Indeed, within ten minutes, Enrique P.'s nurse interrupted rounds with word he was dead.

Because state law requires that every patient dying in a hospital be definitely certified dead by a physician, I broke off rounds and revisited Enrique to pronounce him genuinely dead. Often, such death pronouncements are an annoying, almost laughable formality—as if an M.D. degree is usually necessary to determine if somebody is really dead. But, in Enrique's case, an unexplainable, macabre fascination tempered my usual irritation in doing this chore. Quickly checking him over to be certain he was indeed without vital signs, I was struck by how awful his lifeless body looked. Enrique's was one of those corpses that seemed visibly to decompose from the moment of expiration. Already appearing ancient before death, he seemed to have aged an eternity at the instant of expiration. His skeletal rib cage, bony extremities, and leathery skin reminded me more of a three-thousand-year-old Egyptian mummy than someone who had been alive only a few minutes earlier. His skin, which was never better than an off-color yellow when he was alive, had already turned a dark amber hue within a few minutes of his last breath, and his straggly beard and mustache were caked with dried sputum, snot, and vomit, the latter also matting his tangled locks of hair. His glassy eyes and parched mouth were both halfway open, frozen in mid-breath. Although I knew it was too early for putrefaction to have set in, the pungent smell of hospital disinfectant—305 had just been mopped and cleaned earlier that morning—complemented the stale smells of stool, urine, and vomit, creating a rancid odor of death and decay.

I immediately called his wife with the bad news. Over the telephone she seemed to take it well—at least there was no weeping or wailing—and she said she wanted to come in and see him one last time, before the body was removed to the morgue, a common family request. Thus far, everything still seemed routine: a terminal ex-prisoner with AIDS had just died, and his family had been notified and was on the way to the hospital for a last bedside visit. Soon, Enrique's pathetic remains would be zipped into a hospital body bag and quietly shipped down to the morgue, and 305 and its bed would be scrubbed for the next patient.

A patient's death, whether expected or unexpected, always disrupts the regular floor routine with various tasks that must promptly be done before the body can be released to the funeral director. Even before Enrique's actual death, the nurses had called pastoral care earlier in the morning, when it looked as if he was about to die, and Father Jack had already offered up a prayer and administered Last Rites. When death was definite, I called Sister Pascal, to alert her to his wife's impending visit, knowing that Pascal would want to be there to help comfort her at Enrique's bedside. The nurses were already busy with their required ritual of cleaning up the body for the morgue. At least his wife would not have to see him covered with drying vomit and lying in a puddle of liquid stool. And I began filling out the details of the death certificate and telephoning the city medical examiner, both chores being mindless annoyances that must be done whenever a patient dies. Never are the health-care bureaucrats more concerned about having every minute detail for a patient just right than when the patient is dead.

Fortunately in Enrique's case, the paperwork rituals of death were easy enough. Past experience with such death minutiae helped me toss off the death certificate and complete the call to the city medical examiner in only five minutes. It seemed that the dull record-keeping of Enrique's death was as routine as his hospitalization had been.

Pleased with my efficiency in dealing with Enrique's paperwork, I felt that the only remaining chore was his wife's impending arrival, when I would give her my routine words of reassurance that "he didn't suffer at the end." Having already finished the greater part of rounds on my other patients that morning, I started writing their progress notes at the small desk at the unit's nursing station, which was right across the hall from room 305, where his body was awaiting his wife's final visit. Within a few minutes of starting my chart work, I had already forgotten about Enrique, the most recent death on Unit 3A.

About forty-five minutes later, the security guard at the hospital information desk called the unit to report that Carol P. was there and wanted to visit the floor. Visitors before noon had to get permission to come up, and I verified it was okay. A minute or so later, a well-dressed, middle-aged woman briskly walked off the elevator and went into 305, without stopping to identify herself to anyone. My brief glimpse of her from the nursing station was consistent with the way Carol sounded over the phone: solidly middle class, with a composed and serious bearing. However, her composure quickly shattered after she entered Enrique's room and closed the door.

Immediately, loud shrieks and wails of grief pierced through the door and into the hallway. Knowing it was best to give her some time to mourn alone with the body, I first paged Sister Pascal, to inform her that Enrique's wife had arrived, then I waited five minutes until the crying subsided before going in to comfort her. It still seemed all so routine to me—I had done this many times before.

As I approached 305 and started to open the door, I expected to witness yet another routine deathbed scene between the deceased and the bereaved, with the grief-stricken family member standing mournfully at the bedside, perhaps holding the hand or stroking the head of the dead loved one. Instead, I was jolted out of my complacency by a scene that would be seared into my memory forever. Quietly sobbing tears of loss, Carol was neither standing nor sitting at bedside. Rather, she was lying in the deathbed with Enrique, stretched out on her side, facing him and cradling his head close to hers. Incredibly, she was even tenderly stroking his dirty hair and beard. Never in my many years of observing death had I witnessed such a spectacle: a young woman, who was so much *alive,* was lying in bed next to a *hideous* corpse, which looked even more discolored, shriveled up, and ancient than it did only an hour earlier.

After an initial, awkward pause, I quickly regained my composure, and because she had apparently not heard my entry into the room, I softly introduced myself. Partially turning away from Enrique, she smiled and introduced herself—she was indeed Carol P. Although extending a hand in greeting, she remained close at her husband's side. She choked out thanks for my care, and I in turn recited my standard litany of condolences. Although my words of comfort were well received, I cannot remember exactly what I said to her—I was still trying to keep *my* composure. As I was giving my usual promises that "Enrique didn't suffer at the end," my mind was preoccupied with wondering, "How in the world can she stand to be lying right beside such a frightening-looking corpse?" It took all my professional skill to maintain eye contact with her, rather than self-consciously glancing at the horrendous sight right beside her. Pleasantries finished, I told her that Sister Pascal would be stopping by soon, and I excused myself, leaving Enrique to his wife's embraces.

Back at the nursing station, I mulled over the conflicted feelings I had about what I had just seen in Enrique's room. My initial feeling was revulsion at such a seemingly grotesque sight. But after a few minutes of reflection, my opinion changed from revulsion to admiration and respect.

Enrique's wife must have loved him very much, I concluded, and such unconditional love transcends decaying flesh. What if that terrible-looking corpse in that deathbed were that of my lover or parent or close friend—would not my natural instinct be to climb into bed, to hold and cradle my loved one for one last time, just as Enrique's wife was doing? I realized that her demonstration of love was an image I wanted to remember.

A few minutes later, Pascal arrived on the unit and visited with Carol in 305. Talking with me afterward, Pascal admitted that she, too, had been taken aback at seeing Enrique's wife lying in bed with him.

"I've seen it only one other time," she commented, an observation that, given the amount of death she had seen over the years, validated my own initial shock.

Pascal told me how just yesterday she had read to him at bedside a get-well card his wife had sent him. "Enrique, I love you forever and beyond," it read. Even though he was comatose at the time, Pascal felt obliged to read it to him, just in case he could understand it. Pascal also related to me that Enrique's wife had just told her that she, too, was HIV-positive and had been married to him for ten years, the approximate duration of his prison sentence.

"It's just wonderful, isn't it?" Pascal continued. "Here is this woman who loved him for all these years he was in prison." With characteristic understatement Pascal mused, "A loving heart . . . they tell me this is what counts in the long run."

A few minutes after Pascal left the floor, I received word that the ER was sending up to my unit another ex-inmate who had been medically paroled a few months earlier and who was also in the terminal phase of AIDS. This latest parolee had been assigned to 305, once it was vacated. Still fresh in my mind was the scene of Enrique's wife tenderly holding his emaciated body next to hers, and I realized that there was really no such thing as a "routine" case of an ex-prisoner with AIDS.

Before Enrique P. was finally moved to the morgue that morning, his wife washed his hair and shaved his beard.

• • •

ELVIS C.

Sometimes the needs of a dying patient's family must take precedence over the wishes of the patient, as was the case with Elvis C., who was next to occupy 305.

Elvis was a twenty-six-year-old Hispanic man who began to develop symptoms of AIDS just a few months after he had been paroled from prison. Like many Spellman patients, Elvis first tested HIV-positive when he entered Sing Sing five years earlier, the source of his HIV infection having been either teenage drug use or, in his words, "dirty women," an unseemly macho pejorative often heard from similarly frightened patients. Elvis's AIDS-related illness was central nervous system toxoplasmosis, a destructive parasitic infection that slowly began destroying his brain. Medication temporarily halted the parasite's growth, but other AIDS complications—malnutrition and severe diarrhea—had gradually rendered him weaker and weaker, until he was almost bedridden, requiring help with basic chores such as getting dressed, walking to the bathroom, and sometimes even feeding himself.

Elvis was not the only person in his circle to be stricken by AIDS. He sometimes spoke longingly of his girlfriend—"my girl"—who had died in an AIDS hospice several months earlier, just before he was paroled. His older brother, Fabian, had died the previous year from AIDS and drug-resistant TB, soon after being released from another upstate prison. Fabian had always been Elvis's hero, the big brother whom he always imitated and followed, as if he were pursuing him through the circles of Dante's concentric underworld—all the way to prison and now beyond the pale of AIDS. And, of course, several of Elvis's friends, both from prison and the old Bronx neighborhood, were already dead or else very sick as well.

After he was paroled from Sing Sing, Elvis was far too weak to live alone. Even on a good day, he could walk only a few steps with a cane. He thus stayed with his mother and her invalid eighty-two-year-old mother, in their small fourth-floor walk-up apartment in the Tremont section of the Bronx. A quiet, deferential lady, Elvis's mother spoke no English, despite having lived in America for almost twenty years. Mrs. C. divided her time between a part-time job in a local laundromat and her many home responsibilities, which included caring for her sick mother *and* providing for Elvis's ever-increasing needs. Their one-bedroom apartment was crammed with two large hospital beds, one for her mother in the living room and the other for Elvis in the bedroom. Mrs. C. herself slept on a cot in the kitchen. A visiting nurse looked in on Elvis and his grandmother twice a week, but the major chores devolved on Mrs. C., who kept them clean, helped feed them, took them to their many doctors' appointments, and made sure they took their many medicines.

After Elvis was admitted from the ER to 305 for increasing diarrhea and weakness—he had fallen at home several times the prior week—his mother was able to visit him only a few times a week, since she had to pay a neighbor to watch her mother during the long trip to St. Clare's. Mrs. C. usually brought Elvis home-cooked food, which he never touched, and after tidying up his bed and fluffing his pillow, she would quietly sit at his bedside, not saying much of anything. Elvis, too, would never say much to his mother on these visits, and the two of them would always just stare quietly out into space. Then, an hour or so later, Mrs. C. would get up from her bedside chair, refluff Elvis's pillow, kiss him good-bye, and leave for the Bronx—to resume caring for her mother. Although weak and malnourished, Elvis was still alert enough to talk to his mother during these visits, but he never did. Once, this monotonous bedside scene was enlivened when, to Elvis's obvious delight, his mother brought in from home his older brother's black motorcycle jacket. Elvis grinned ear to ear as he tried on Fabian's jacket for the first time, and thereafter he would proudly wear it over his hospital gown. In mute testimony to past and future transgressions by Mrs. C.'s two sons, on the back of the jacket a skull and crossbones had been burned over the words PERDONAME, MADRE—"Forgive me, Mother."

As Elvis's condition worsened at St. Clare's, it was apparent to everyone that he would require nursing home care—apparent to everyone, that is, except Elvis. Over the first week of his hospitalization, he became so weak that he required constant help feeding and washing himself, and he could no longer take even a few steps on his own. Even under the most ideal conditions, home care at his mother's would have been a daunting project, requiring twenty-four-hour nurses and his mother's continuous attention. Abby A. met several times with Elvis and his mother to discuss where he should go once he was released from the hospital. Through an interpreter, Mrs. C. said she very much wanted her son to return home, but she was worried about how she could handle both Elvis and her mother, who already took up most of her time. Furthermore, these discussions revealed that Mrs. C.'s health was not the best either—she had heart problems and severe arthritis that limited how much she could do. Thus, it was felt that the additional work of caring for Elvis at home would be too much for Mrs. C., a conclusion she herself agreed with, albeit reluctantly. Mrs. C. would have to send either her mother or her son to a nursing home in order to care for the other in her home.

Empathizing with the terrible choice confronting Mrs. C., the ward staff and I tried to discuss the matter with Elvis, in hope he would understand his mother's dilemma and acquiesce to going to a hospice nursing home located only a few blocks from his mother's apartment. It was repeatedly emphasized that once he was in the hospice, his mother could visit him daily, not just twice a week. Elvis, however, simply did not understand how sick he was and how difficult his home care would be. A lifetime of being indulged by his mother had convinced him it was his right to expect her to look after him at home, especially since he was sick and since—as he always reminded everyone—his mother had been able to care for Fabian during the few months before his death. Repeated admonitions and pleadings from nurses, social workers, and other staff failed to budge Elvis from his determination to go home. He would simply exasperate everyone by cheerfully dismissing their entreaties with his often repeated declaration that "my mama will take care of me." Perhaps it was a combination of selfishness and HIV dementia, but his simpleminded, carefree attitude on the subject was impenetrable to reason, threats, or pleas. Soon, he became totally oblivious to discussion of the matter and would smile back with the vacuous grimace of advanced HIV brain disease.

During these discussions with Elvis, his mother would often be present, and even though a translator would apprise her of what was being said, she would just quietly sit or stand by the bedside, looking torn and helpless. She was not able to tell her son that he could not come home, that she simply was not physically able to care for him any longer. More than once during these debates with Elvis, her thoughts probably drifted back to twenty years ago, when she left Puerto Rico for the mainland with her mother and two boys in tow, her sons named after her rock-and-roll idols. Never on that fateful trip had it ever occurred to her that she would someday bury her sons before she buried her own mother. Despite her optimism, their pilgrimage to the promised land had taken a terrible turn. As if the violent procession of drugs, violence, and prison had not been enough to wreak emotional devastation, there was the final, crushing blow of AIDS, a monstrosity totally incomprehensible to her, which laid an absolute, uncontested claim to her two sons' lives. Mrs. C. had not had an easy life. Uneducated, she had had to take two or three menial jobs at a time, just to feed her boys, who by default had only the neighborhood gangs to parent them in the Bronx. Mrs. C. had much to think about as she would keep her silent watch in room 305, at Elvis's bedside.

At last, a decision could no longer be postponed. Elvis had stabilized to the point that continued hospitalization was no longer feasible, and given the infectious risks of staying in the hospital, it was really not in his best interests to remain in St. Clare's. Mrs. C. had neither made preparations to take Elvis home nor told him point-blank he had to go to a nursing home. Moreover, the nursing home a few blocks from their apartment was about to open a hospice bed for Elvis, a bed that would be quickly snatched up by someone else if he did not take it right away. The situation seemed at an impasse: Elvis would not budge from his insistence on going home, and his mother could not tell him she could not take him home.

As his doctor, I knew my primary allegiance was to Elvis, but I also recognized that his fate was inextricably intertwined with his mother's and, indirectly, with his invalid grandmother's, whom I had not even met. Moreover, if Elvis went to his mother's, it was quite likely her inability to give appropriate care would cause him—and his grandmother—to deteriorate faster. There was another reality as well: Elvis could no longer stay at St. Clare's, and by then totally dependent on others for his care, he really had no choice in the matter. Thus, as soon as the hospice bed became available, I asked Abby A. to arrange ambulance transportation for the next morning and to notify his mother about the planned transfer.

Sometimes, my job is to assume the role of the bad guy and lay out the unpleasant facts to stubborn patients such as Elvis. So, when the ambulance crew arrived on the ward the next day, I accompanied them to his room, where his mother was already at the bedside, still saying nothing, but appearing especially conflicted. Upon seeing the ambulance crew with their stretcher, Elvis knew what was about to happen. The day before, I had told him that he was going to be discharged to the hospice, but as always, he brushed aside my news with oblivious nonchalance, saying, "My mama will take care of me." But now they were coming, literally, to take him away—and in front of his mama, who was *supposed* to take him home.

Elvis immediately frowned and reaching toward his mother began to whine, "No! No! *Casa! Casa!*" Her son's cries instantly prompted her own silent tears and trembling, and she fumbled through her old purse for her handkerchief.

I realized that Elvis's frantic fear provided me one last chance to shake him into reality. Summoning all the mystical authority my age and doctor's white coat could evoke, I leaned over him, looked right in his face, and forcefully told him he had no choice but to go to the nursing home that day. He was too weak to resist, he had to leave the hospital, his mother—much as she wished to the contrary—could not take him home, the only place he could go was the nursing home, and if he really loved his mother, he would relent and go to the nursing home in peace.

"Look at her!" I angrily commanded, my outstretched arm pointing to his mother for emphasis, while I maintained eye-to-eye contact with him. *"This is tearing her apart!* She loves you dearly, but she can't take you home. For once in your life, show her you love her and stop putting her through this hell!"

The stridency in my voice jolted Elvis, who had always been accustomed to a pleasant demeanor from his doctor. Appearing terrified and remorseful, he started again to cry, this time with deeper, more mournful sobs—tears that had little to do with the nursing home matter and more to do with a lifetime of unspoken regrets, unresolved losses. Wiping away his tears with my handkerchief, I caressed his head and softly told him he would be all right, that I knew he really loved his mother and wanted to do the right thing for her. Throughout this emotional confrontation and catharsis, one of the nurses stood beside Mrs. C. and translated first my rebukes and then my reassurances. Yet, she still said nothing to her son and stared down at the floor, quietly sniffing back tears. She appeared to be weighed down by a sorrow to which words could provide no dimension or form.

After a minute or so, Elvis calmed down and actually helped the attendants and nurses with his transfer to the ambulance stretcher. While he was being moved, his mother quickly gathered up his few possessions—his brother's motorcycle jacket, the brown teddy bear that had belonged to his "girl," the small crucifix over the head of his bed, a few dated motorcycle magazines, and some old family snapshots taken many years ago on a family outing to City Island.

As the ambulance crew wheeled Elvis out of 305 and down the hall toward the elevator, his mother and I silently followed behind. In the past, whenever I had asked her if she had any questions about her son's case, she would never have any. For the last time, I asked, through the nurse interpreter, if she had any last questions or concerns she wanted to ask me.

Mrs. C. looked up at me and, with great weariness and even greater love in her voice, quietly replied, "I have in my heart the pain that only a mother can feel. Thank you, Doctor, for taking care of my last son." The eloquence of her words froze my heart in my throat.

Then Mrs. C., Elvis, and the ambulance crew boarded the ward elevator. Already back to his oblivious baseline, Elvis was smiling—he was holding his girl's teddy bear in his arms—and he waved good-bye to the gathered ward staff as the elevator door closed.

Three weeks later, Abby A. told me she had heard that Elvis C. had recently died in the hospice nursing home. And I felt a small part of the pain that only a mother can feel.

• • •

ANGEL T.

The one word that best describes Angel T. in her living and her dying is *class.* Indeed, Angel was probably the most colorful patient ever to stay in room 305, a veritable extraterrestrial being compared to Enrique P. and Elvis C., her two immediate predecessors.

Right up to her death at the age of thirty-one, Angel T., perhaps like a Brooke Astor clone gone awry, was the grande dame of a rarefied, elite society—the largely secret, ethereal subculture of Hispanic, Asian, and African-American transvestites and transsexuals, who defined much of the flamboyance of the New York night world in the 1980s. Although to outsiders Angel's lifestyle seemed alien, to Angel and her friends it was the only *natural* way to live. Her world was immortalized in the cult film *Paris Is Burning,* which actually starred Angel and her attendants. Resplendent with sequined gowns, outrageous fashion shows, and tons of makeup, Angel's world was concomitantly real and fantastic, both tragic and glamorous.

The quintessential, defining event in this largely invisible society was the weekly fashion show held in one of the city's underground private clubs, where shimmering armies of sleek drag queens vied against one another for attention and adulation, preening and strutting a mimelike dance, referred to as voguing. The denizens of Angel's subculture were a mixed lot. Some were merely transvestites, others were their butch admirers, and a few were bona fide transsexuals. Angel herself was a "partial transsexual": her genitalia were still male, her breasts had been rendered female by hormones and silicone, and her psyche was decidedly molded by *Vogue* and *Cosmopolitan* magazines. The most serious devotees of this fantasyland lived together in "houses," where they felt sheltered from the harsh opprobrium of the outside world. There they enjoyed the heady freedom of communal life with other drag queens and transsexuals—the freedom of sharing makeup tips or just vacuuming the carpet in nylons and heels, whenever the mood would hit them. As testimony to her verve, her style, and her genuinely good heart,

Angel had long been "mother" of her house, the House of Chenise, which was preeminent among them all. Indeed, the house's sisters were always the best-dressed and classiest drag queens at the voguing shows, in their way on par with the top Paris runway models.

Over the years, Angel had shepherded many younger initiates through the rigorous rituals of eyeliner, lipstick, and nail polish. More importantly, she provided these novices with reassurance and validation that, yes, it was okay just to be yourself, even if being yourself meant lip-synching to Madonna while slithering around in outré outfits Madonna would covet. To be sure, the more baleful side of this special sorority of fashion mavens was never far away. Indeed, there was always the lurking danger of drugs, alcohol, and—tragically—frequent, almost obligatory suicides that resulted from the loneliness of living a lifestyle condemned by family and society at large. But, despite the travails of living a life of feminine sensibility in a man's body, there were scant tears—after all, that would cause the makeup to run. Smiling on the outside, crying on the inside—that was the credo of any drag queen worth her eyeliner.

Angel had, in fact, a lot to cry about, but she rarely did. Ironically, Angel's major AIDS complication was Kaposi's sarcoma, the terrible cancer that often first appears on the skin but can later infest and destroy practically any organ. Angel's KS caused a progressively severe disfigurement of her legs, arms, and face. Born with a classic, feminine beauty that professional models would envy, Angel had always been on the petite side, with graceful legs and small, soft hands. Her facial features had likewise appeared delicate and somehow eternally youthful, and her high-pitched, childlike voice—reminiscent of Michael Jackson's—endowed her with an air of fragility, tenderness, and vulnerability. Angel's personality perfectly matched her baby-doll appearance: never bitchy or demanding, she was all sweetness and grace up to the very end. And the heartbreaking end came painfully slowly.

Angel's KS became increasingly disfiguring and grotesque. As the tumor multiplied unchecked, her legs and arms ballooned to many times their normal size, the purple skin thickening and wrinkling into a rough, nodular hide that would break open and ulcerate into hideous, seeping sores. Angel's classic, angular facial lines were obliterated by similar tumors; only the winsome, doelike eyes remained miraculously spared. Her hands became so bloated that eventually she was no longer able to hold her combs or makeup brushes, much less eating utensils.

But try as it might, Angel's KS could not rob her of her endemic beauty and style. Whereas to an outsider Angel might have looked like a mutant monster out of a science fiction movie, to her friends her radiant beauty remained undefiled. Day after day, a steady stream of her girlfriends from the house visited her at bedside—not to mope or grieve, but rather to cheerfully assist their mother with her daily beauty needs. The first challenge to Angel's friends was the *dreadfully* dreary atmosphere of room 305, which *absolutely horrified* her sisters on their initial visit. Rolling up their blouse sleeves, Angel's girlfriends scrubbed every inch of the room to a high polish. Like ladies-in-waiting, they brought in new drapes, large soft pillows, cozy quilted blankets, vases of fresh flowers, and several large throw rugs. Angel's small pink vanity table, brought in from the house, was set up in a corner of the room, to hold her best wigs and several framed pictures of her in her heyday, when her beauty was at its peak.

Keeping Angel's beauty at that peak was her friends' second mission, and as befits drag queens from the House of Chenise, they were more than up to the task. Armed with bulging vanity cases and an assortment of wigs and colorful, if not always sedate, lingerie, Angel's angels daily painted her nails, applied facial makeup over the KS lesions, assisted her with lipstick and eyeliner, and adjusted her wig to the right fit. The ubiquitous deformities of her KS were minor blemishes in the eyes of her friends. Surrounded by such solicitous devotees, Angel seemed more like an eternal goddess—a Venus with attendants or Diana Ross before a concert—than a dying AIDS patient.

Indeed, this daily hospital ritual had long been familiar to Angel, who over the years had likewise waited on so many of her sisters who had been afflicted with AIDS and gone on before her. Angel was perhaps among the last of a dying breed: AIDS had already decimated the voguing houses to the point that Angel and her girlfriends seemed quietly resigned to their inevitable passing, much like Judy Garland or Marilyn Monroe, tragic figures ineluctably drawn to a tragic end. Although her attendants were never mournful in their daily ministrations, this homage somehow seemed evocative of the final days of another great beauty, Cleopatra, and her few remaining maidens, as they waited in royal majesty for the inevitable arrival of the Roman army and, with it, death. In Angel's case, however, the fatal sting of AIDS, and not an asp, would strike her down.

Nonetheless, Angel and her sisters knew that too much self-pity was unbecoming to a drag queen. As she once waggishly said of her illness, "It's *only* AIDS, honey—not something *really* serious like a run in your nylons right before you go onstage!"

Once Angel was properly fixed up, she and her girlfriends would settle back to chat about those important things all girlfriends focus on—their love lives, the latest gossip, the new fall fashions, how all men were pigs and sluts, and the upcoming round of weekend parties. Tears or sad words were not allowed, as if each meeting were like a bridal shower. AIDS and KS were politely ignored or, more accurately, snubbed. Throughout her illness—and the last days were truly horrendous—Angel maintained a calm serenity, a quiet dignity that only the word *class* could do justice to. That great New York socialite of an earlier era, Lady Astor, stoically awaiting her fate on the deck of the listing *Titanic,* had absolutely nothing on Angel T., preparing for her fate in her deathbed in room 305. Both doomed ladies had an elemental sense of good taste, an innate sense of how to behave when adversity hit: in a word, class.

This touch of inordinate class saw Angel through to the end. As her lungs filled up with fluid from the KS and as her breathing became increasingly labored, she never flinched. She never wavered from the supreme self-assurance that had guided her through the shoals of a partial transsexual's life. After all, if she could successfully navigate herself and her sisters through the hatred and derision their lifestyle evoked, then she could certainly handle a mere trifle like AIDS. And as for the criticism that she and her friends had used their fantasy world of high fashion and voguing parties as escapism, as an easy way out of dealing with AIDS, Angel would probably have retorted, undoubtedly with a slightly bitchy edge, that until there was a cure for AIDS, then makeup and lipstick—not to mention a nice set of heels, a sequined evening gown, and matching wig—could carry the day as well as toxic drugs such as AZT or DDI.

Angel's wake and funeral were held in Spanish Harlem, at the Holy Zion Pentecostal Roman Catholic Church of the Madonna. The survivors of the New York voguing scene, surrounded by their retinues and curious hangers-on, were in attendance, bedecked in their black mourning gowns, gray wigs, broad-rimmed black hats with veils, extravagant plumes of feathers, and seemingly miles of pearls. New York had probably never witnessed a funeral with more unadulterated camp. It was an open-casket funeral, of course, and throughout the calling hours and services—right up to the final closure of the lid—her girlfriends were still in attendance at casket-side, applying a tad more lipstick, adjusting her wig, fixing her eye shadow and liner just right. Angel wore her favorite low-cut turquoise gown, complete with her ostrich-feather boa and rhinestone tiara. The casket seemed to float, enveloped by a backdrop of dozens of elaborate floral arrangements, while off to the side several large, framed photographs of Angel in happier days, posing in full regalia, complemented the spectacle. A mixture of gospel hymns, torch songs, and Madonna served as background music.

As with Judy Garland on view at the Campbell Funeral Home, a long line of friends and well-wishers filed past the casket, some pausing to kiss her hand, others putting into the casket small mementos such as photographs, rosary beads, rings, and flowers. There was no wailing or loud sobbing—only quiet weeping. Angel would of course have wanted it that way, since, after all, too many tears cause the makeup to run.

• • •

MABEL C.

A few Spellman patients have had family members with whom I have been unable, despite my best attempts, to establish the emotional connection necessary to help everyone—patient, family, and myself—face the patient's death. Mabel C., the next admission to 305, was such a patient, whom I have always remembered not only for her unique AIDS story, but also for her daughter Ana, who steadfastly refused my help in dealing with her mother's illness. Such refusals have challenged me to look beyond my anger and hurt feelings and to accept my limitations as a physician.

Mabel C. was not a typical AIDS patient, which, of course, meant she was a typical *Spellman* AIDS patient.

Having emigrated from Costa Rica many years earlier, Mabel was already in her early fifties when she was infected with HIV. She had never had a blood transfusion and had never ever dreamed of shooting up with drugs. Rather, Mabel's only risk factor for HIV had been a remote monogamous relationship with a younger man who, unknown to her at the time, enjoyed recreational IV drugs. When this ex-boyfriend died from AIDS several years later, Mabel's two grown daughters prevailed upon her to be HIV-tested. Never a strong person, Mabel simply could not cope with testing positive and tried to kill herself by jumping in front of a subway train. Misjudging the distance of the oncoming train—her cataracts ultimately saved her—she jumped onto the tracks too soon and was rescued by an off-duty cop. Immediately thereafter, Mabel was admitted to Bellevue's psychiatric unit for a psychotic depression requiring heavy-duty medication and several subsequent psychiatric hospitalizations before she could live alone again. Once her mental illness stabilized, her older daughter, Ana, enrolled her in the Spellman outpatient clinic for care of her HIV disease.

Mabel remained asymptomatic from her HIV disease for several years, but because of her chronic depression, she often missed many of her Spellman Clinic appointments. About one year before her death, Mabel had her first St. Clare's hospitalization on 3A. By then sixty-three, she was one of Spellman's older patients, but she was not unique in that regard, since more than a few of Spellman's patients were in their sixties, and even more were already in their fifties. Older AIDS patients usually handle complications of HIV disease much more poorly than their younger counterparts, but the reason for Mabel's first Spellman hospitalization was completely non-HIV related. She merely had congestive heart failure, a common affliction for many people her age, as well as worsening of her psychotic depression. Indeed, her T-cell count then was around 500, well above the 200 level at which opportunistic infections begin to threaten.

Lasting six weeks, Mabel's first hospitalization—a full year before her death during her second hospital stay in room 305—was a protracted and vexing affair for two reasons. First, Mabel's generalized debility and psychiatric illness made her recovery painstakingly slow, and second, her older daughter's demanding, stubborn personality alienated me to the point that I stopped trying to help her deal with her mother's illness. Mabel and her medical problems, these I could deal with—but her daughter Ana was the difficult part of the Mabel C. case.

Mabel herself was a short, obese lady who was always content to stew away in bed rather than get up and care for herself. This physical deconditioning was further worsened by her congestive heart failure and psychotic depression. This latter problem was deceptive, for on cursory evaluation, Mabel seemed to be a pleasant, cheerful lady who projected an engaging sweetness. However, on closer questioning, she would cheerfully admit that she regularly heard voices telling her to "take a knife and *kill, kill, kill!*" Other times the voices would command her to push someone off a subway platform in front of a train or to slit her own wrists. Never violent in demeanor, Mabel was nevertheless placed on constant nursing observation until antipsychotic drugs eradicated the voices and their chilling commands. Yet, despite the improvement in her heart failure and psychosis, she remained quite frail. She needed nursing help with even basic daily activities such as dressing and bathing, and she remained mentally fragile enough to require close supervision in taking her medicines.

For several weeks during that first hospitalization it was uncertain when, if ever, Mabel would be ready to return home to her apartment in the Bronx. Moreover, if she ever did become strong enough to go home, twenty-four-hour nursing care would definitely have to be arranged. As her doctor, I had to consider seriously the option of discharging her to a nursing home, especially if she did not get much better. Of course, Mabel herself wanted to return to her apartment, but she really did not understand the extent of her dependency on others.

Complicating the decision about what to do with Mabel was her strong-willed older daughter, Ana, whose stubbornness, in retrospect, turned out to be right for Mabel, at least for the subsequent last year of her life. An attractive, prosperous-appearing young woman with an intense personality, Ana projected a "superior creature" persona, dressing the part with stylish, expensive clothes accentuated by Gucci scarfs, shoes from Bergdorf's, and Ferragamo purses. Her brisk, to-the-point manner contrasted with the slow, befuddled responses of her mother, whose modest clothing and short, squat appearance further made them an unlikely mother-daughter pair. An executive in a Madison Avenue advertising agency, Ana visited Mabel almost every day in the hospital, and although never questioning her mother's medical care per se, she always complained to me about things over which I had no control—the cold food, the chronic shortage of nursing staff to help feed Mabel, the malfunctioning hospital bed, the infrequent cleaning of her room, the lack of a shower or bath in her small bathroom, the peeling paint on the bathroom ceiling, the occasional cockroach scurrying across the floor. Normally I would sympathize with such complaints, since I, too, was disgusted with these and many other affronts to my patients.

Moreover, whenever I would caution Ana about her mother's precarious condition during that first hospitalization, Ana would always appear slightly skeptical, ever so slightly annoyed. It was as if she not only did not believe me but also did not want to hear it in the first place. The skepticism I could handle—over the years I have often had to deal with distrust from patients and their families. But her silent annoyance with what I was telling her—a perturbed, condescending air of "Why are you telling me this?" impatiently conveyed with arched eyebrows, withering gaze, and tightly pursed lips—*really* upset me and eventually made me hate talking with her about Mabel.

Perhaps my feelings about Ana were colored by frequent complaints from Mabel's nurses about how Ana would never help her mother during her visits. She would never feed Mabel, get her things like tissues or water, or give the nurses an occasional helping hand, which most other visitors would give under similar circumstances. Expecting the nurses to behave as servants rather than caregivers, Ana always exuded an air of superiority, of entitlement for both her mother and herself, as if they were on the Gold Coast floor of Lenox Hill Hospital.

I always came away from conversations with Ana feeling angry and frustrated. I realized that some family members, for reasons private to them, simply do not want emotional support—that I should not let my hurt feelings color my reactions to Mabel. But Ana's rejections of such emotional support always seemed to have an extra edge, an extra bite of gratuitous contempt that eventually made me avoid talking with her. Thus, as Mabel's first hospitalization dragged on, I confined my discussions with Ana to just the cold facts about Mabel's case, without expressions of concern or empathy. Of course, Mabel herself was always a joy to visit; she was happy and pleasant, despite the voices still sometimes telling her to "kill, kill, kill." Whereas during that first hospital stay I held Mabel's hand, both literally and figuratively, I eventually gave up trying to hold Ana's.

After five weeks, Mabel improved to the point that, with assistance, she could get about her room and care for herself. The terrible voices had left her, and her heart failure stabilized, although she still required oxygen for her breathlessness. Plans for Mabel's discharge—either to a nursing home or to her apartment with home care—had to be finalized soon. Ana, of course, would hear nothing about sending her mother to a nursing home: it was out of the question, period, end of discussion. She had no problem with twenty-four-hour nursing care in Mabel's apartment and thought it was a good idea, given Mabel's frailty. However, Ana adamantly drew the line at any suggestion that *she* be involved in her mother's home care. Helping the visiting home-care nurses was simply out of the question, since, as Ana frequently reminded everyone, she had a "very busy and very important job" that precluded such commitment to Mabel's home care.

The only problem with Ana's insistence on noninvolvement in Mabel's home care was that all of the city's visiting nurse services had always stipulated that, as a condition of their services, there be available a friend or family member to provide emergency backup care in the event an unforeseen emergency prevented the nurse from making it to the patient's apartment. It was rare for such emergencies to occur, but subway problems, sudden illness affecting the visiting nurse, or a host of other developments could leave Mabel stranded alone at home for hours, hours during which she could injure herself or become sick. Such a designated emergency backup person had to sign a contract with the visiting-nurse agency, promising to be available if needed. Since Mabel had no close friends and since Ana was the only family member living in New York—her sister lived in Texas—this backup had to be her.

As the time for a decision approached, Abby A. and I had the temerity to tell Ana that the choice was either sending Mabel to a nursing home or agreeing to sign the visiting-nurse contract, and she bristled at the ultimatum, icily informing us that she would "go all the way to the top" of the administration of St. Clare's and the visiting-nurse service: "I do not have time for this nonsense . . . no one dictates to *me.*" But Ana got nowhere with her protests to the powers at "the top," and she grudgingly signed the visiting-nurse contract, promising to help her mother if an emergency arose.

Soon thereafter, Mabel was discharged to her Bronx apartment, with twenty-four-hour nursing care. Because she was so fragile, both physically and mentally, I was skeptical about her chances of staying at home for very long and expected her to bounce back into the hospital in a month or so. During the debate about Mabel's discharge plans, I had often expressed to Ana these reservations regarding home care, and Ana dismissed my concerns. I have always been overly pessimistic and overly protective toward my patients in similar situations, often forgetting that human beings, even when sick, are much tougher than I give them credit for. And Ana proved me wrong in my cautious worries. For almost a year after her first hospitalization, Mabel remained stable, quietly perking along at home under the supervision of her visiting nurses. And as far as I knew, Ana never had to be inconvenienced with emergency backup duty.

Visiting me monthly in my Friday-afternoon outpatient clinic—the home-care nurses now made sure she kept her appointments—Mabel was her cheerful, noncomplaining self. Although the nurse accompanying her to the clinic would always have a list of complaints Mabel had apparently voiced at home, Mabel herself would always deny or minimize these symptoms when I would ask her about them in the clinic. The threatening voices remained at bay—the nurses likewise saw to it she took her psychiatric medicines—and her congestive heart failure was fairly well controlled on the home oxygen and the multiple heart and blood-pressure medications she was taking. Mabel's T cells still hovered in the 400-to-500 range, and she refused my advice to start AZT or DDI. "I'm on enough pills," she would reply, a statement I had to agree with. Indeed, except for being HIV-positive, Mabel was very much like the myriad other "little old ladies" with high blood pressure and heart disease from my years of private practice in rural Iowa. Sedentary and even more obese, Mabel existed a heartbeat away from catastrophe. Never on any of these clinic visits was Mabel accompanied by Ana.

Mabel enjoyed a good year in her home because of her constant nursing supervision *and* her daughter's original insistence that she not be sent to a nursing home. However, Ana's stubbornness unfortunately did not serve her mother as well at the end of her life.

A year after her first hospitalization at St. Clare's, Mabel was brought to my Friday-afternoon clinic as an unscheduled emergency visit, this time accompanied by both her nurse and Ana, the latter's presence a tip-off that something was seriously wrong. Earlier that morning Mabel had complained to her nurse about a severe headache and then developed slurred speech and trouble walking, even with assistance. Through thick, garbled speech, Mabel, who appeared more lethargic than usual, told me in the clinic she was "just fine" and, as always, denied having any symptoms whatsoever. But my exam in the clinic confirmed my fears: Mabel was having a stroke, which was already fairly extensive. Her entire left side was paralyzed.

Leaving Mabel in the exam room with her nurse, I stepped outside into the noisy, crowded clinic hallway with Ana, to explain to her my diagnosis and my concerns—that Mabel's stroke was severe and could very well progress further. If the STAT head CT scan I was about to order showed a blood clot in Mabel's brain, I explained to Ana, then little could be done except for bed rest and to hope that the paralysis would resolve with time. I further cautioned that the complications of a stroke in someone as frail as Mabel could be serious: pneumonia, blood clots in the lungs, blood infections, plus a host of other potential problems. I did not want to alarm Ana, but I felt I was obligated to apprise her how serious Mabel's situation was.

However, while I was telling Ana my concerns about Mabel's stroke, I once again got that sickening feeling, as I picked up the same put-out, it's-your-problem-not-mine, why-are-you-telling-me-this facial expression that caused me a year ago to walk away from further attempts at empathy and emotional support. But the gravity of Mabel's new problem forced me to continue and bring up a matter I hesitated to raise but had to: CPR/DNR. If Mabel were to deteriorate to a cardiopulmonary arrest, I asked Ana, should resuscitation be attempted or not, especially given her frail condition and the risk of her lingering on in intensive care before finally dying. Ana's impatient, annoyed mien intensified to barely concealed hostility when I worried out loud that Mabel might not make it this time.

"Why are you talking to me about this when as far as we know she may get better like she did the last time?" she snapped, her voice laced with resentment and indignation, as if I were wishing a fatal outcome on Mabel.

Normally, at such a tense juncture with a distraught family member, I would still press on a little more, to explain why I had to raise such an emotional issue. I would elaborate upon the New York State law regarding mandatory CPR, even if hope of meaningful recovery were small, all the while trying to give the necessary emotional support and empathy. But Ana's comments were just too much for me, and remembering the failure of my previous attempts to communicate with her, I decided not to persevere and gave up, but without the self-pity I had a year earlier. I switched to my businesslike, just-the-facts mode. With blank expression and monotone voice, I reviewed Mabel's diagnosis, the diagnostic tests that would be done, and my therapeutic plan, as meager as it was. Maintaining an uncharacteristic manner of cold, clinical detachment, I matter-of-factly concluded by pronouncing Mabel's prognosis "very poor and very guarded." Having no questions, Ana sullenly returned to her mother in the exam room.

I had to realize, I later thought to myself, that some people simply do not want emotional support, at least from the doctor. *My* feelings should not be the issue, I had to tell myself. Rather, I would assume the role of her mother's "body mechanic," tending to Mabel's tired old bones and—for Ana at least—sparing the expressions of concern and empathy. Of course, I felt bad for Mabel and feared that she might have to pay for her daughter's contentious refusals to face her dismal prognosis, as well as pay for her doctor's exasperated surrender to her stubborn daughter.

I made arrangements for Mabel's admission to my ward—to room 305—and, before leaving the clinic, stopped by the exam room to see her one more time. Ana had apparently just left for her job, and only the nurse was there. Mabel was still awake and comfortable, blissfully unaware of how terribly sick she was and how annoyed her daughter and doctor had been with one another only a few minutes earlier. Holding her paralyzed hand and patting her on the shoulder, I lied and reassured her that everything would be okay.

Late that night, Mabel lapsed into a coma—probably from progression of her stroke—and soon thereafter she had a cardiac arrest. A code was called, the nurses and on-call intern did the obligatory CPR, and—amazingly—

Mabel was snatched from the grave and was transferred to ICU, where she was hooked up to a breathing machine. By the time I saw her on rounds the next day, she was still comatose, hanging by a sliver of her brain stem. Her vital signs, largely controlled by the lower brain stem, were dangerously erratic, with irregular and rapid heart rate, a body temperature wildly fluctuating between 92 and 105 degrees, and a barely measurable blood pressure, which was requiring higher and higher doses of intravenous "presser" drugs for support. Mabel was not breathing on her own, the breathing machine next to her bed dutifully pumping her lungs with oxygen-enriched air. Predictably, her body had already been violated by the usual, almost ritualistic trappings of "intensive care"—every body orifice had been probed or tubed, and both arms, including her paralyzed left one, had been tied down to keep her from pulling out her various tubes.

Nonetheless, despite being enveloped in the shroud of high-tech medical devices, Mabel appeared strangely at peace, as if her soul had already departed her body—that what remained was content to passively endure the machinations of intensive care. Unlike other critically ill patients, Mabel was not fighting against the breathing machine, but rather was completely motionless save for the regular, periodic heaving and sighing of her chest, as the respirator breathed in and out for her. Her eyes were closed, and her face was almost beatific. Mabel was, mercifully, brain-dead: both the stroke and the shock of the cardiac arrest had shut off the oxygen supply to her brain for so long that most of it was now dead, except for the deepest regions of her brain stem, which still tenuously controlled her vital signs. It would be only a very short time before the thread of brain stem suspending her over the grave would snap.

Satisfied that Mabel was feeling no pain, I left the room in search of Ana, but the ICU nurses reported she had visited earlier in the morning and had left word for me to telephone her at work. After being transferred through several receptionists, I finally reached her and explained Mabel's condition in the most candid terms: that Mabel's vital signs were unstable despite aggressive treatment, that Mabel was comatose and not suffering any pain, and that it was unlikely that all these heroic measures in ICU would save her. Responding with characteristic aloofness, Ana did not focus on Mabel's well-being, but was more intent on pinpointing who was to blame for the cardiac arrest.

"Why wasn't someone watching her more closely?" she demanded to know.

The entitled, annoyed tone in her voice over the telephone made it clear that she would not accept my explanation that the floor nurses could not be in Mabel's room every minute. Nor did she sound interested in my reassurances that her mother was not in any pain, and she definitely did not appreciate my reminder of how I had cautioned her about Mabel's guarded condition the day before.

"She was *definitely* not that sick yesterday," she shot back.

For Ana, her mother's medical care was always someone else's responsibility, someone else's job. Abruptly, but courteously, I reverted to my detached, antiseptic persona, ending the telephone conversation with perfunctory remarks about how "everything possible" would be done to keep Mabel alive. The full weight of "heroic" intensive care and its high-tech medical hardware should be sufficient to deflect any further criticism from Ana about Mabel's medical care, I thought as I hung up the phone.

"Everything possible," of course, was not enough for Mabel: she died later that same evening from another cardiac arrest. Her ticker just stopped and could not be restarted, despite thirty minutes of chest pumping by the ICU staff, multiple electric shocks to her chest, and pints of emergency cardiac medications pushed into her IVs. The doctor in the ICU called Ana to inform her of Mabel's death, which I did not learn about until the next day. I never heard from Ana after that and do not know how she took the news. Normally in such situations I would contact the family the next day, to express my condolences and to answer any questions or concerns they might have. However, in Mabel C.'s case, I decided there was no use attempting such niceties with Ana. Nonetheless, I was both sad and happy for Mabel; although I would miss her, I would always remember her look of eternal peace in the ICU.

At the next Spellman memorial service, to which all family members are routinely invited, Ana was not in attendance. I would like to think that, for Ana, compassion and grief for her sick mother were simply too personal to share with anyone, even with her mother's doctor. If such were not the case, then Ana really died long before her mother did.

• • •

JUDITH A.

Sometimes a Spellman patient's death can be a scary experience—not for the patient, but for me.

No sooner was Mabel C. urgently transferred to ICU and room 305 hastily cleaned, than the ER sent up Judith A. to take her place. Judith's final hospitalization was for worsening *Pneumocystis carinii* pneumonia and had been preceded over the prior two months by four or five brief hospital stays for the same problem. On each of these previous admissions, Judith would initially come in with severe shortness of breath, would improve after a few days of intravenous Bactrim, and by the fifth or sixth day would sign out of the hospital, not completing the necessary twenty-one-day course of treatment for her pneumonia. For Judith, hospitalization at St. Clare's was an inconvenience, an annoying hiatus that kept her from the only passion in her life—crack cocaine.

Forty-four years old and addicted to crack longer than she could remember, Judith was inseparable from Leroy, her boyfriend. An intensely muscular man weighing nearly three hundred pounds, all of it muscle, Leroy had an intimidating way about him, but his intense stare was what especially disturbed me whenever I would talk with him. During all of Judith's hospitalizations he was constantly at her side, virtually dominating all discussions I would try to have with her about her medical care. From the moment she would be admitted to the moment she would leave against medical advice, Leroy was there, and with a seriousness that bordered on paranoia, he would hang on my every word, my every contact with Judith on rounds.

Judith always let Leroy do the talking. She never appeared afraid of him, but seemed totally overwhelmed by his energy, his strength, and his obsessive love and concern for her. It seemed that Judith was Leroy's entire life, his reason for being. From his perspective, she *had* to live, if not forever, then at least for a long, long time. And no one could ever tell him otherwise.

Each subsequent time Judith was readmitted for pneumonia, her condition would be a little worse. She would be a little weaker, a little more malnourished, a little more short of breath. Although part of the reason for this deterioration was her partially treated pneumonia, there was also lung damage from cocaine, causing a type of emphysema called crack lung, which further compromised her lungs' ability to fight off the *Pneumocystis* organisms. It would be only a short time before Judith's lungs failed and she would die, especially if she persisted in her crack use and her penchant for using St. Clare's for "drive-through" medical care.

Despite her drug use, Judith seemed to understand what she was risking by always leaving the hospital before completing treatment of her pneumonia.

"I know you're right, Doc, but I gotta do what I gotta do, even if it kills me," she would matter-of-factly answer my repeated warnings.

Leroy, on the other hand, did not have a clue about how serious Judith's condition was. Each time she would be readmitted, I would try to prepare him for the worst by explaining that Judith was very sick, that there was no guarantee she would get better, and—this was the thing that perplexed him the most—that her worsening pneumonia could kill her. I was not trying to scare Leroy, but because of his overwhelming concern about Judith, I felt I had to give graphic assessments about her prognosis. Leroy never wanted to hear what I would try to tell him.

Whenever I would privately raise this unpleasant possibility with him, he would first just look at me with a dumb expression of disbelief and anger. "She's not *that* sick, is she?" he would ask, less as a question than as a command or statement of fact. When I would counter that, yes, Judith *was* that sick, he would always look at me with the same intense, almost menacing stare. I felt as if I were prey being sized up by a predator. Immediately I would try to extricate myself from this stalemate by carefully reexplaining to him Judith's illness and prognosis, emphasizing how everyone was doing everything possible to help her, how her prematurely leaving the hospital compromised her medical care, and how sometimes the best possible medical care was not enough to save someone. As I was reviewing these issues with him, Leroy would shift his stare from me to his feet and shake his head in the negative, as if he could neither hear nor understand me. If I would try—very gingerly—to penetrate this reverie of denial and press him to acknowledge that Judith was indeed very sick, he would break off his trance, look up from the ground, and once again just silently glare at me.

This silence was uncharacteristic for Leroy. Although he never backed up my periodic warnings to Judith about her cocaine dependency and her leaving before her treatment was completed, he had often interjected his opinions about Judith's medical treatment, criticizing everything from her dirty room to her nursing care. The only time Leroy seemed to distance himself from Judith was when the subject of her drug use would come up. "I don't know . . . I just don't know," he would passively respond in a detached monotone to my periodically urging him to help Judith lay off the cocaine.

I could never make any emotional contact—any human connection or link—with this angry man. I grew to dread, and fear, any contact with him and hoped that, if Judith were to die at St. Clare's during one of her hospitalizations, it would not be on my unit, not on my watch. The thought of having to tell Leroy that Judith was dead gave me a gnawing, sick feeling in the pit of my stomach.

Although it was always aggravating whenever Judith would use St. Clare's as a revolving door, I was never angry or upset with her. She was a pleasant enough lady, and unlike many crack addicts, she never became abusive or irritable as she withdrew from cocaine. Except for her once telling me she had three teenage children, who never visited her in the hospital, I never really got to know her well—both her brief hospital stays and Leroy's hovering, domineering presence saw to that. All of the staff on her Spellman ward would repeatedly try to help her with her addiction to coke, but the best efforts of the psychiatrist, social worker, and pastoral care staff never made any difference.

During one of Judith's prior admissions, a month or so before her final stay at St. Clare's, I had raised the matter of cardiopulmonary resuscitation and DNR with her and Leroy, and I quickly learned not to raise it again. Before she could even reply, Leroy jumped in with the first and last word on the issue: "That would be *murder!*" he adamantly retorted about the option of signing a DNR, his voice trembling with agitation and—to my unease—great menace. "We want *everything possible* done to keep her alive, *no matter what,*" he insisted, more as a threat than a request. Judith quietly demurred to Leroy's wishes.

By her final hospitalization, Judith had lost almost thirty pounds, and her breathlessness had progressed to the point she needed oxygen just to walk to the bathroom in 305. Her almost continuous, raspy cough made conversation difficult, a development that did not matter for her, since Leroy, her constant companion and mouthpiece, was at her side. Her chest X ray showed a progressive worsening of the pneumonia, and her arterial blood gases, while still acceptable, were not as good as those from previous admissions. Leroy could just not understand why Judith was getting worse and worse from one hospitalization to another.

"She's not getting better . . . are you giving her the right medicine? Maybe the nurses are giving the wrong pills. Why isn't the medicine working?" he demanded of me, oblivious to the cause-and-effect relationship between Judith's deterioration and her never following through with her medical treatment, let alone her crack use. By the time of this final hospitalization, I had ceased any attempts at explanation or reassurance. I only hoped that on this admission, like a cat that would always land on its feet, Judith would once again rally and improve enough to return to the streets and her beloved cocaine.

The second night of her final hospitalization, at a little past 4 A.M., the nurse found Judith on the floor next to her bed, unresponsive and without vital signs. A code was called, and she was successfully resuscitated and shipped off to intensive care, ironically to the same bed just vacated by Mabel C., who had expired a few hours earlier. In ICU, Judith was promptly hooked up to a breathing machine. During the resuscitation, a nurse found three empty crack vials in her bedsheets.

By the time I arrived at the ICU later that morning, the doctors there had already set in place "everything possible"—to use Leroy's words—to keep Judith alive. Although the resuscitation had restarted her heart, her brain was not as fortunate. Judith was comatose, and her breathing was totally dependent on the breathing machine next to her bed. Multiple antibiotic medications—for both PCP and bacterial pneumonia—were pouring into her several IV lines, and enveloping her flaccid body was the usual array of high-tech monitoring devices.

I quickly reviewed her chart and examined her, to be sure I agreed with the treatment begun by the ICU doctors. Judith's brain had probably been destroyed by lack of oxygen from her unwitnessed cardiopulmonary arrest five hours earlier. No one really knew how long she had been "dead" before the nurse found her, but it must have been considerably longer than the crucial three minutes the brain can survive without oxygen. Now a gooey mush that would never recover, Judith's brain was gone forever, along with its memories, as well as more mundane functions such as thinking, moving, and breathing. Just as with Mabel C., the empty shell of Judith's body was now hanging by its brain stem, which itself was becoming unstable: her body temperature and blood pressure were fluctuating wildly. At least Judith had no pain, I thought to myself as I went through the motions of examining her, all the while dreading my pending stop at the ICU waiting room, where, the nurses had already informed me, Leroy was waiting to see me. How would he react? What could or should I say to him, especially after my many unsuccessful attempts to prepare him for this eventuality? For the first time in my several years at Spellman, I feared for my personal safety, and I decided not to get cornered by Leroy in the ICU waiting room.

No sooner did I appear at the door of the waiting room—positioning myself right next to the doorway for quick exit—no sooner did Leroy see me than he jumped to his feet and, more angry and distraught than I had ever seen him, demanded of me, "What the fuck happened last night! It has to be somebody's fault—I want to know who fucked up!" His intense stare seemed even more penetrating, more malevolent than usual. He was almost trembling, with anger and remorse seething from every perspiring pore.

"It has to be somebody's fault!" he repeated, this time, curiously, more as reassurance to himself than as a statement of fact.

Carefully and calmly, I explained that there was no way to know for sure what had happened to Judith. Perhaps her pneumonia or asthma had suddenly worsened, perhaps she had developed a blood infection, perhaps an unexpected heart attack or blood clot in her lungs had occurred—no one could say for sure. My voice was breaking. To my relief, Leroy's countenance suddenly changed from anger to a perplexed remorse, his thick brows now furrowed with agitated worry.

"She's going to be all right, isn't she?" he insisted of me.

Again, calmly, I shook my head in the negative and replied that, although "everything possible" was being done to save Judith, I was "very worried" that she might not make it, hoping that my empathetic worry might divert some of his anger away from me. Leroy seemed amazed at my pessimism.

"But aren't you doing *everything possible*"—and here he seemed to imbue the words *everything possible* with a reverential, almost mystical power—"aren't you doing *everything possible* to help her?" he pressed me, sounding genuinely incredulous at my words.

Right then I realized the depth of Leroy's bewilderment. He appeared totally dumbfounded when I proceeded to reveal to him a crucial fact of life—namely, that our loved ones will die someday and that sometimes everything possible might not be enough to keep them from dying. Maybe it was the popular myth about medical science's infallibility, or perhaps his own immaturity, but Leroy was convinced that death would not happen to Judith. By now more annoyed than angry, he returned to his mantra that Judith's crash in the night had to be somebody's—*anybody's*—fault.

Then, even more carefully, I mentioned the crack vials found in Judith's bed, feeling emboldened enough to subtly remind Leroy that maybe her cocaine addiction was at fault. Instead of exploding, he suddenly became very still, quickly averted his eyes from mine, and stared intently downward, shaking his head in the negative and inaudibly muttering to himself. It seemed that something terrible was tormenting him. Entranced in this reverie, Leroy did not respond to my asking if he had any further questions, and I was happy to excuse myself and leave him to his thoughts.

Over the next few days it became apparent that Judith would never awaken from her coma. Several EEGs—brain-wave tests—returned pathetically flat line, without the slightest trace of higher cerebral activity. Judith was brain-dead, that frequent, macabre result of a "successful" cardiopulmonary resuscitation. Moreover, subsequent urine toxicology results returned positive for cocaine, confirming my suspicions that, sadly, Judith had essentially done it to herself, that if there was any "fault" in her catastrophic demise at 4 A.M. several nights earlier, it was hers.

However, after many years of "blaming" patients for their illnesses, I had long ago realized that such blame games only make the doctor feel superior, to the detriment of patient care. Most patients do not realize how doctors all too often secretly blame patients for their illnesses, smugly commenting about patients' smoking, obesity, sexual practices, noncompliance with medications, drug use, personality disorders, or a host of other "lifestyle" causes of patients' illnesses. In reality, all the blame game does is assuage the doctor's frustrations with the crude imperfections of medical science. Blaming the patient is, in fact, an excuse that hides the doctor's powerlessness over most disease, an especially vexing impotence for those doctors who are fixated with issues of control. What doctors who play the blame game really hate most is those patients who have no bad habits, who follow the doctor's advice to the letter, and who *still* get sicker and sicker, thereby leaving the doctor accountable for some other explanation for medical science's failure to cure. With age, experience, and self-knowledge, a good doctor recognizes his or her own frailties and imperfections and becomes tolerant—indeed loving—about patients' frailties and imperfections.

After Judith's transfer to ICU, my daily visits with Leroy remained troubling and fruitless, since, in his own way, he was emotionally "brain-dead," at least as far as understanding Judith's extreme condition was concerned. Efforts by the Spellman pastoral care staff were likewise unsuccessful.

"Leroy just doesn't get it, does he," an equally frustrated Sister Pascal remarked after several attempts to help him face the inevitability of Judith's death. "He obviously loves her very much, but I sense there's something else that's causing the pent-up anger."

Pascal's suspicions confirmed my own impression that he was tormented by more than just Judith's condition. Whatever it was, Leroy was still insistent on blaming me—and everyone else who had been involved in her care. Every day, he and I had the same inconclusive "conversation" the two of us had had that first day in the ICU waiting room, each of us talking past the other, never connecting on any emotional level. Leroy was overwhelmed by an abiding obsession with affixing fault for Judith's cardiopulmonary arrest and ensuing coma.

"It's *got* to be somebody's fault for what happened," he would repeat day after day, as if just saying it made it true.

Six days after her transfer to ICU, on a Friday at 10 A.M., Judith legally died: her heart gave out and prolonged CPR failed to restart it. Leroy had just arrived for his daily visit when the CPR began, and Sister Pascal had ushered him into the ICU waiting room, while the ultimately futile resuscitation attempt played itself out. After pronouncing Judith dead, I warily headed toward the waiting room, not quite knowing what to expect when I delivered the grim news. Fortunately, Pascal was there with Leroy, but the apprehension in her face confirmed what I already knew—namely, that the two of us would be no match for Leroy if he were to lose control.

As soon as I entered the small waiting room, Leroy stopped his nervous pacing and fixed his eyes on me with a piercing intensity that somehow transcended all of his other stares.

With an almost inaudible whisper I announced, "I'm sorry, but Judith is dead—her heart gave out . . . we couldn't get it started again . . ." My voice trailed off as I briefly paused to gauge his reaction.

His stare remained frozen on me, but with an uncharacteristic glimmer; there was the slightest break in the intensity that somehow signaled disbelief. Yet he said nothing, remaining totally motionless. It was as if he was still processing the tragic news.

"We did everything humanly possible to save her," I tentatively stammered on, "but her pneumonia was too far gone for us to pull her through."

Leroy still said nothing, but a trembling agitation was slowly building up throughout his body. His eyes began to squint and blink faster, his brows started to furrow and knit spasmodically, and his upper body began to sway slightly from side to side. He was still silent, but thankfully his intense stare gradually dissolved into tears—indeed, soul-cleansing tears, I thought to myself. His whole body started to shake and convulse in bone-deep spasms of tearful remorse, and I reached out and gently touched his upper arm, making my very first contact with this troubled man's heart.

"I know you loved Judith so very much." I wanted him to know that I did understand that his obsession with Judith had been based on love. Tears now overflowed his eyes, dripping off his cheeks, onto his chest and the floor. Giving him my handkerchief, I continued to stand beside him, my hand still on his shoulder.

"It's all my fault, it's all my fault," Leroy choked out through the tears. "I gave it to her, I gave it to her, it's my fault, I shouldn't have done it, I did it . . ."

After giving Leroy a few minutes to cry, I quietly resumed the standard reassurances I had always given a loved one: that Judith had felt no pain and did not suffer at the end, that the hospital staff shared his grief, and that his presence was a wonderful comfort to Judith throughout her illness. But, just as with my prior conversations, Leroy did not appear to hear me. Now more composed, but still snorting back tears, he just sat on the sofa in the waiting room, staring at the floor, repeating over and over again to himself, "It's all my fault, I gave it to her . . ."

Leaving him to grieve further with Sister Pascal, I pondered his repeated lament "I gave it to her." I wondered to myself about the "it" that was torturing Leroy. Was "it" the cocaine she took that night at 4 A.M., or was "it" the HIV itself?

Leroy never revealed his torment and never returned to St. Clare's thereafter.

• • •

JAY M.

Jay M. was the last resident of room 305 during the three months preceding the quarterly Spellman memorial service. Unlike his six predecessors, Jay's name was *not* listed in the Book of Remembrance in that service, or in any of the subsequent services.

As an affluent gay white male who had never used drugs—let alone been inside a prison or in a bread line—Jay M. was not a typical Spellman AIDS patient, but his memorable affection for both the staff and more typical patients on his ward denoted a courage that was typical of many Spellman patients.

Gay white males, especially those with jobs and health insurance, have always been a rare species at Spellman, and once admitted to St. Clare's, they are often unhappy with the amenities or, more accurately, the lack thereof. Most of these atypical Spellman patients—as well as their families and friends—have been aghast at the filth, noise, and overall disarray. The result has been depressingly predictable: both an embittered patient who vows never to return to St. Clare's again, and frustrated doctors and nurses who are weary from all the complaints about the problems over which they have little or no control. Years of such unpleasant confrontations have soured feelings on both sides. Many of the city's gay-oriented AIDS advocates have had little regard for St. Clare's and its AIDS patients, and not a few veteran Spellman staff silently recoil in apprehension whenever a middle-class gay white male lands there. These prejudices are, of course, regrettable, since Spellman's typical AIDS patients are as deserving of advocacy as gay white male patients are deserving of clean, quiet hospital rooms. Jay M., with his good humor and empathy for others, defied this stereotype. In so doing, he earned the affection of everyone—staff and patients—who knew him.

Struggling to catch his breath that first night of his first hospitalization, Jay ended up at St. Clare's the same random way most other gay white males before him had arrived—namely, by city ambulance, which must bring any unstable patient to the nearest hospital emergency room. Indeed, St. Clare's was just three blocks from Jay's Midtown apartment. A thirty-one-year-old teacher for mentally retarded children, Jay had ignored the insidious breathlessness and dry, hacking cough that had begun several weeks earlier. He was afraid to see his private doctor, and only a few days before his arrival at St. Clare's was he HIV-tested for the first time, at an anonymous testing site in the Village. The positive result turned his fear into despair, further paralyzing him from seeking medical attention for his worsening symptoms.

A neighbor called 911 when Jay, returning from the grocery store one evening, was unable to catch his breath after climbing two flights to his walk-up apartment, and he was rushed to St. Clare's. A chest X ray in the ER showed his right lung was partially collapsed, and in quick sequence, the on-call surgeon inserted a tube into his chest, his collapsed lung reexpanded, and he was shipped up to Unit 3A, still uncomfortable but breathing much easier than before. The medical intern who saw Jay late that night correctly suspected *Pneumocystis carinii* pneumonia, PCP, and started him on high-dose intravenous Bactrim, oxygen, and steroids, the latter medication to decrease lung damage from the PCP. By the time I first saw Jay later that first morning, he was much more comfortable.

Moderately balding and lean to the point of appearing scrawny, Jay looked tired from the previous night's ordeal. His thin frame did not really appear emaciated—there were still muscles and a little fat on his bones—and he really reminded me more of the typical ninety-seven-pound weakling than a malnourished Spellman AIDS patient. Jay's apprehension seemed perfectly congruous with his slight body build, but beyond the worry in his face, I sensed a strength of spirit that I hoped would help him through whatever travails awaited him. Though softly trembling, his voice evoked respect. During my initial exam, Jay related to me some of his life—his teaching job, his fears about his illness, his close-knit family in Arizona, and his moving to New York both to help inner-city children with learning disabilities and to live openly as a gay man. Here, I thought to myself, was a good-hearted person with AIDS who also happened to be gay. I really wanted him to make it, as did the rest of the 3A staff.

During that first admission, tests confirmed that Jay indeed had PCP, and he quickly responded to therapy, regaining his strength and putting on ten pounds. By the second week, the chest tube had been pulled, and he was fully ambulatory, even starting light exercises in his room.

Although Jay's prompt response to anti-PCP therapy was not unusual—with appropriate treatment, most patients with first-time PCP will pull through—his emotional response to St. Clare's was decidedly unusual. Despite the glaring shortcomings of my AIDS ward, Jay never voiced strident or inappropriate complaints. Whenever he reported legitimate gripes, such as problems with his menu orders or lack of heat in his room, he never acted as if it were my fault. Nor did he ever seem to mind rubbing shoulders with the less well-scrubbed patients on the unit. And, not surprising to him or to me, most of his ward neighbors—the ex-prisoners, the active or former drug users, the homeless—all responded to this friendly, easygoing white man with the same civility and respect he accorded them. Jay somehow understood the limitations inherent in AIDS care. Moreover, he seemed to appreciate how most AIDS patients have much more in common than what demographics might indicate.

Despite the routine hardships of being a patient, Jay liked his Spellman floor because he liked me and the other staff there. To his nurses and physician assistants, he was the ideal patient. He never refused his medications, he always allowed blood work and X rays, he never used illicit drugs, and he made every effort to help himself rather than sit back and expect to be waited on. Jay was one of those rare people who looked for goodness in everyone, and if he did not find it, he tried to cultivate it in them. His sweetness and good humor were infectious: even surly radiology technicians and blood drawers from the lab were charmed.

Jay's parents visited him from Arizona several times during that first hospitalization. Investment bankers who managed their own firm in Phoenix, they were every gay man's dream parents, not only loving and concerned, but also matter-of-factly accepting of their son's homosexuality. Their emotional support was all the more remarkable since, prior to Jay's illness from AIDS, they had not even known for sure if their son was gay. Mr. and Mrs. M. were also every attending physician's dream parents: they fully understood my explanations to them and did not second-guess me by insisting—or even suggesting—he be transferred to a more renowned AIDS center. Moreover, like Jay, Mr. and Mrs. M. even appeared to sympathize with the impossible job I had as an AIDS doctor at St. Clare's. Indeed, it had taken me many years of being a doctor to understand that I should never expect the patient or family to empathize with, or make allowances for, whatever difficulties I might have in providing medical care. In a word, Mr. and Mrs. M. wanted the best for their son *and* were perceptive enough to sense that, despite the drab rooms and crumbling infrastructure, Jay was already getting "the best."

Jay's first hospitalization was uneventful, and within a week of completion of treatment for his PCP, he was back at work, teaching at his school in Brooklyn. For the next eleven months, he did well, coming in to see me for clinic visits every month or so. As with his inpatient stay at Spellman, he always stuck out a little while waiting in the Spellman Clinic for his appointment. Well-dressed, deferential, and not nodding off from methadone or illicit drugs, he appeared more suited for a waiting room of a private medical practice. But Jay came to the Spellman Clinic because he trusted the care he relieved there—it was *his* hospital, and I was *his* doctor. I had come to St. Clare's to care for the indigent AIDS patients there, knowing there were plenty of gay and gay-sensitive physicians in New York willing and able to care for patients like Jay M. But his clinic visits were nonetheless a change from patients whose primary concerns were food stamps and finding a crime-free homeless shelter.

Unfortunately, as the first anniversary of his AIDS diagnosis approached, Jay began to have night sweats and occasional low-grade fevers, along with an insidious fatigue that limited his work to six hours a day. Soon thereafter, at a clinic visit, I noticed on his lower legs a few small, raised purple spots, which he said had popped up a few weeks earlier. Jay seemed worried when I casually advised biopsy of one of these spots. He already knew Kaposi's sarcoma, the AIDS cancer, was a possibility. I tried to be reassuring, even though I, too, shared his worry.

Our fears were quickly confirmed by the skin biopsy: Jay now had KS. He took the news fairly well, trying to be bravely blasé—"I guess I have to die of something, so it might as well be this"—but a hesitancy was also in his voice. A few days later, his worried parents called from Arizona, wondering what this news portended. Although I could assure them that Jay was okay for now, I could not predict how aggressive or quiescent Jay's KS would be. KS is notoriously variable in its severity and progression, some patients perking along indefinitely with their tumors, and others rapidly succumbing to widespread tumor invasion of almost every organ system.

During the ensuing weeks, Jay became more easily exhausted and had to go on disability leave at his job. His low-grade fevers were supplanted by bona fide high-grade fevers, with drenching, enervating night sweats. Then quickly came the dreaded harbingers of real sorrows for a person with AIDS—a dry cough and subtle breathlessness. A year after his first admission to St. Clare's, Jay was very sick again, and when he saw me on an emergency visit in the Spellman Clinic, I tried to hide my alarm that, instead of a recurrence of PCP—which could probably again be treated successfully—his new respiratory symptoms might signify spread of KS to his lungs, a rapidly fatal complication of AIDS. Jay was likewise very worried that first day of his second hospitalization—he, too, knew he was very sick—but his fear was tempered by the trust and affection he had for me and the other friends he had on the Spellman staff: "You guys got me through the last time, so I know you can do it again."

Passing largely in the confines of room 305, Jay's second, and probably last, St. Clare's hospitalization spanned almost four weeks over the Christmas holidays. Not only did he have a pneumonia, but his lungs on chest X ray were studded with multiple golf-ball-sized masses, an unusual and ominous presentation for any patient, regardless of his HIV status. Among the many possibilities to consider, at the top of the list were KS and several other malignancies, all of them incurable and with dismal prognoses. There was only the most slender of hopes that these lung masses were rare, unusual forms of TB, fungus, or even rarer bacterial infections—the slim hope being that such less likely diagnoses might at least be treatable.

The search for Jay's diagnosis—in the hope that a treatable illness could be found—comprised one of the more spectacular wild-goose chases in the annals of the Spellman Center. For, despite weeks of extensive evaluation and tests—including three bronchoscopies, one needle biopsy of the lung, and one open-lung biopsy under general anesthesia—the diagnosis of his lung masses remained frustratingly elusive. Several pulmonary consultants, pathologists, and infectious-disease specialists examined, reexamined, and re-reexamined his chest X rays, chest CT scans, and multiple biopsy specimens. Some of these experts thought the tumors were KS or lymphoma, others favored bizarre fungal balls, and others were not sure what to think. Experts from St.Vincent's Hospital were called in, again to no avail. Jay's lung problem was undiagnosable, at least short of a postmortem exam. Even then, there was no guarantee that an autopsy would be any more conclusive than the gaggle of tests and procedures that had already been done. Although many AIDS-related diagnoses can be difficult to make, it was nevertheless highly unusual for so many AIDS experts to be so baffled.

Throughout this diagnostic ordeal, Jay kept up a good face, but beneath the veneer there lurked genuine terror. The frustrating lack of a diagnosis, despite the many tests done on him, never shook his trust in me or the Spellman Center. As on his first hospitalization, he always appeared understanding and would always try to put *me* at ease when I would have to tell him the latest disappointing news.

"I know you're trying your best, Doctor Baxter . . . we'll get through this okay, I just know we will." His cracking voice betrayed him.

As the weeks of fruitless search for a diagnosis passed, Jay became thinner, his weight dipping down to one hundred pounds. This weight loss, his overall fatigued appearance, and his worsening baldness made him look many years beyond his age of thirty-two. Although still able to walk, he now required a cane and had the gait of an old man; only the boyish smile and sparkle in his eyes recalled the lost youthfulness of the recent past. Shuttling between Phoenix and New York, Mr. and Mrs. M. visited Jay many times during this four-week hospitalization, one alternating with the other every three or four days. Their son was slowly dying before them.

Despite the specialists' promises that the next, more invasive biopsy would definitely clinch the diagnosis, everything came back negative—or ambiguous enough to confuse the so-called experts. Probably the ultimate disappointment, the proverbial last straw for everyone, was when an open-lung biopsy, done under general anesthesia and with direct visualization of Jay's lung masses by the thoracic surgeon, turned up results that a small army of the best pathologists in the city could not agree upon. Some argued for KS, others for lymphoma, others for fungal infection, and—the final insult—others for a repeat open-lung biopsy!

By that point in the ordeal, both Jay's increasing debility and the lack of any diagnosis made it pretty clear to everyone that Jay was probably going to die soon from an undiagnosed or undiagnosable lung complication of AIDS. Jay and his parents needed no convincing that it was no longer worthwhile to fight. Although never hostile, his weary parents were as perturbed as I was, and they only wanted him home. Jay's comfort for the presumably imminent end was now the only objective for everyone.

The planning for Jay's discharge was easy, since money was no concern. After setting up home nursing care and arranging transfer of his case to an internist in Phoenix, his parents and I simply set a discharge date, and they flew back together to New York to pick up their son. A few days earlier, they had hired a moving company to pack up the things in his studio apartment, and by the day of discharge, the moving van had already made it halfway to Arizona. Having only been in New York for a few years, Jay did not have any close friends in the city, other than "work friends" at his school, and thus the move back home to the desert Southwest would not be difficult for him. After all, his parents were his entire life at this point—they were all he needed. Fortunately, Jay was still able to take care of himself and to walk short distances, so he and his family could share some time together during the last few weeks to months of life everyone felt he had left.

The morning of discharge, Jay was more garrulous than usual. He probably realized he would not be returning to St. Clare's and his friends on Unit 3A; rather, he was leaving for good, to die in Arizona. Having arrived early that morning, his parents had brought in an assortment of doughnuts, bagels, juice, and coffee for the staff, and Jay himself had already distributed to other patients on the unit the flowers, magazines, paperback mysteries, and junk food that had accumulated in room 305 over the prior weeks. No sooner did I arrive on the unit than Jay and his parents were packed and ready to leave for the car waiting downstairs to take them to the airport. All of the discharge paperwork and prescriptions had been taken care of the day before, so there would be no delay. Pausing at the nursing station for final good-byes, Jay and his parents looked as if they were embarking on a vacation rather than a final trip home to wait for death. The Louis Vuitton suitcases were neatly stacked on a carry-on cart, and coats and lighter bags were draped over their arms and shoulders. Mr. and Mrs. M. thanked the assembled nurses, PAs, and myself for our care. Jay was trying to hold back the tears.

"Can I call to tell you how I'm doing at home?" he asked me.

I was barely able to get out my affirmative reply before he continued on, haltingly and self-consciously, "And . . . and do you want my folks to call you"—here his voice cracked slightly, but not enough to tarnish his courage—"to call you when I die?"

Jay's seemingly innocuous question—whether I wanted to know when he died—encompassed universal concerns all of us face: namely, the significance of memory as a form of immortality, the question of who will remember us when we are gone, and more sadly, whether such remembrance really matters anyway. Both Jay and I knew, either consciously or subconsciously, that the question he had asked epitomized what it meant to be human: to care about—that is, to hold in memory—another human being while we are alive, and to hope for the immortality of remembrance when we are dead.

Part of being a good doctor is knowing how to touch a patient when words alone are insufficient. As a good-bye to Jay, a handshake would have maintained a proper, professional distance, and a pat on the shoulder would have been too timid and wimpy. Rather, what the situation called for was a big hug, which Jay, frail as he was, returned with a strength that surprised me.

"Of course I want to know," I whispered to him.

Then Jay and his parents walked onto the elevator and left 3A. I wondered if I would ever hear from him again. I honestly believed I was saying good-bye for the last time.

Almost a month later, Jay called me at the hospital. Firmly settled in his parents' home in Arizona, he seemed to be doing remarkably well, at least for him. He was strong enough to go to movies, to eat out with some old high school friends, and to go shopping with his parents. His twenty-four-hour home-care nurse was no longer necessary. His doctors at the University of Arizona, who had received his medical records and substantial X-ray files from St. Clare's, were likewise clueless as to what had been going on in his lungs, but—and here I did not believe my ears—Jay said that a repeat chest X ray done there had shown that the multiple lung masses were now completely gone. Perhaps, I thought, Jay was just confused, or perhaps his doctors dissembled a bit to give him enough hope to enjoy his remaining weeks to months without worrying about the lung tumors.

That Jay was able to go out and be more active did not surprise me too much. Like dying stars that explode into supernovas before collapsing into darkness, many terminal patients will briefly rally before their final nosedive to death, especially when they are out of the hospital and back in familiar surroundings. Indeed, the pleasant Arizona climate after leaving a gray New York winter could be a temporary tonic for even an end-stage AIDS patient. Genuinely delighted to hear from him, I gave Jay words of encouragement, updated him on the latest news and gossip at Spellman, and asked him to remember me to his parents. Jay especially wanted me to give his hellos to the 3A staff.

It was a brief conversation, five minutes at most, and after hanging up, I once again recalled his question to me a month earlier. I anticipated a sadness in the telephone call I would someday receive—soon, I thought—from his parents, announcing his death. There was no way, I reasoned, Jay had more than a month or so of life left. Those lung tumors, though undiagnosed, were an unequivocal kiss of death, and not even the Arizona sun and Jay's good heart could halt their inexorable spread.

Two months later, Jay called again from Arizona, to fill me in on his condition. He had returned earlier that day from a camping trip in the mountains with his older brother and sister-in-law. He admitted he was "a little tired out," but still felt up to joining his parents later that evening for a movie and dinner out. His voice sounded the same, not any weaker than it did when I last spoke with him, and his spirits were still bounding. He was looking forward to visiting his grandma in Texas the following week. And, yes, the doctors reported that his chest X ray was still clear of any tumors. This second telephone call from Jay—at least three months after he left St. Clare's to die—seemed almost miraculous: Jay M. should have been dead by then.

And so it has been for fifteen months since Jay walked off Unit 3A with his parents, ostensibly to die the death his doctor and all the distinguished specialists had felt was so imminent.

Whenever Jay phones from Arizona—and the calls still come every two to three months—his greetings are like salutations from another world. Always coming when I least expect them—out of the blue, as it were—the phone calls from him gratify and amaze me: *gratify* because my former patient is still alive, and *amaze* because the indomitability of his spirit has confounded the odds and the experts.

Every time we close our phone conversation, I wonder if it will be our last. Jay has gradually become weaker over these many months, and his activities have had to be more curtailed, an indication his disease is still present. But regardless of when he eventually dies, Jay M. has already attained a measure of immortality.

8

IMMINENT MORTALITY: "IS THAT ALL THERE IS?"

"Juan H."...
"Evaristo R."...

Called out from the altar only seconds apart, these two names conjure for me two universes of diametrically opposed memories. For although Juan H. and Evaristo R. were alike in many respects, the haunting contrasts between Juan's fear and Evaristo's courage in the face of death provided striking lessons in dying and living—memorable lessons that compelled me to examine my own anxieties about death. The revelations I gained from these two unlikely teachers will reassure me when I face my own imminent mortality.

BOTH JUAN and Evaristo were state prisoners and were in the terminal phases of AIDS when they were admitted for the final time to the Spellman service, both over the same Columbus Day weekend. Both men were in their early thirties and had contracted HIV from IV drugs, which neither of them had used for many years. "Gave up drugs years ago," Evaristo ruefully noted on his admission interview, his voice edged with uncharacteristic self-pity about how sins of a decade earlier had now laid claim to his life. Evaristo was originally from Puerto Rico, Juan from Colombia, and at the time of admission, both men had been on work release, employed in the community during the day, but obliged to return to prison in the evening. By yet another coincidence, both Evaristo and Juan occupied adjacent hospital rooms on the Spellman service, although from the outset both were far too sick even to wander outside and cross paths. Despite these similarities,

Evaristo and Juan had emotional reactions that were profoundly different to their deaths-in-slow-motion.

While still in prison, before being put on work release, Juan had already had several complications of AIDS, including PCP and CMV retinitis. Although untreated CMV could progress to total blindness, Juan had repeatedly refused on these prior hospitalizations to allow insertion of a large intravenous chest catheter, a Groshong cathether, to administer daily medication to keep the CMV in check. Always evasive as to why he refused such sight-saving therapy, Juan had fortunately not lost further vision by his final Spellman admission, which was for worsening diarrhea. Having lost forty pounds, down to an alarming ninety pounds, Juan had become unable to do the janitorial job assigned to him, and his work release was thus about to be revoked. State law mandated that work-release prisoners too sick to work had to go back to prison, regardless of their circumstances. However, before the Department of Corrections could ship him back to Sing Sing, a high fever, plus increasing fecal incontinence, gained him the relative sanctuary of St. Clare's.

Even on the first day of his final hospitalization, Juan's malnutrition was already profound. His face was gaunt, with sunken eyes and hollowed-out cheeks, and his arms and legs were draped with useless strips of what were once thick, firm muscles. For Juan, this severe emaciation was yet another reason for getting a large intravenous chest catheter. High-calorie IV fluids could be infused through such a catheter, which could also then be used for ganciclovir to treat his CMV retinitis. As his physician, I repeatedly urged Juan to agree to the chest catheter, but as on prior hospitalizations, he adamantly refused.

A few days after admission, preliminary stool and urine tests indicated he had either TB or a closely related bacterium called MAI, which can cause wasting and diarrhea, especially of the magnitude Juan was experiencing. He was immediately begun on a complicated regimen of seven antibiotics that would treat both TB and MAI, but these drugs severely nauseated him, further limiting his already marginal food intake. Even more alarming was the appearance of jaundice only a few days later, possibly from liver damage secondary to the antibiotics. Tragically, Juan quickly found himself in a classic Spellman-style catch-22: he needed prompt treatment for his bowel infection, which was caused by either TB or MAI, but the necessary drugs were making him nauseated and were damaging his liver. Since the offending medications had to be stopped, Juan was considerably worse off than when he had come in only one week earlier. He still had debilitating diarrhea, plus now hepatitis, which was weakening him further.

At the time he was admitted to the room beside Juan's, Evaristo seemed to be in much better shape than his neighbor. Unlike Juan, Evaristo had never had any prior hospitalizations, despite his T-cell count of 10, and except for thrush, he had suffered no serious complications of his HIV disease. A robust-appearing 172 pounds, he had admission symptoms that might seem minor: a low-grade fever, a nagging dry cough, and mild breathlessness whenever he would walk more than a few blocks.

"I just get tireder and tireder," Evaristo remarked about his work-release assignment on street repairs.

Because he had not been on Bactrim prophylaxis against PCP, I was concerned that these relatively mild symptoms might have signified early PCP, but I also noted on his exam some disquieting findings: several small purple growths on his arms, as well as a similar but more reddish growth on his right lower eyelid. Throughout the AIDS epidemic, Kaposi's sarcoma had been the first and best-known AIDS-related malignancy, and although it had been thought to afflict primarily HIV-positive gay men, the experience on the Spellman service had sometimes contradicted this, since women and heterosexual intravenous drug users had also occasionally contracted KS. The seemingly innocuous growths on Evaristo's skin and eyelid raised the specter of this cancer disseminated throughout his body, especially in his lungs, a possibility that was the most dreaded of all KS complications, with rapid deterioration from worsening breathlessness. The patient with lung KS gradually suffocates to death. Evaristo's chest X ray on admission was certainly abnormal, showing an extensive pneumonia, but as is often the case in AIDS, this X-ray abnormality was nonspecific and could have been due to a dozen or so possibilities.

To help with diagnosis, Evaristo underwent a bronchoscopy the day after admission; a pulmonary specialist inserted a flexible scope down his windpipe into his lungs. The results were worse than I had expected: Evaristo had both PCP and widespread KS of the lungs. Because of my initial concern about PCP, I had already started Evaristo on high-dose intravenous Bactrim, which was continued, and Spellman's consulting hematologist-oncologist promptly saw him and started intravenous chemotherapy, to retard progression of the lung KS. Unfortunately, there was no known cure for KS.

Thus, soon after admission to neighboring rooms, Juan and Evaristo knew that they were at the edge, that death was palpably close. But their contrasting reactions to death's imminence produced cautionary, sobering revelations on my rounds.

From the onset of his hospitalization, Juan was scared to death of dying. Despite my repeated explanations about his medical condition, he was unable or unwilling to understand what was happening to him. The only thing he knew was that he was dying. He seemed totally overwhelmed by everything: his increasing weakness, his intractable diarrhea, his enervating nausea, his worsening jaundice—everything about his condition seemed monstrously incomprehensible to him. From the very first day of hospitalization and for every day thereafter, his bulging eyes seemed permanently frozen in a hideous stare of terror, and his eyebrows would be furrowed over his sunken orbits in an expression of fathomless worry. His face always seemed transfixed with a terrible combination of worry and fright.

None of my usual attempts to give Juan emotional support penetrated this obsessive fear. The touch of my hand, my reassurances that I would keep him out of pain, my proddings to get him to talk about his life in hope of learning how to better comfort him—none of these approaches eased his constant fear. Just as he had previously refused to tell me why he would not agree to a chest catheter, he steadfastly refused to discuss any feelings about his illness, repeatedly denying he was depressed and, accordingly, refusing to take the sedatives and antidepressants I had prescribed. Juan's daily high fevers, which always prostrated him, filled him with particular dread.

"What's happening," he would nervously ask me whenever his fever would spike to 104 degrees, *"am I dying?"*

But, despite his desperate inquiries, Juan would never listen to my explanations about what I thought was going on, my repeated reassurances never penetrating his impregnable mask of fear and foreboding. Juan remained absolutely inconsolable: peace of mind seemed as elusive and unattainable to him as a cure for AIDS itself.

Unlike many of his fellow patients at Spellman, Juan had a devoted mother who visited him daily, but her hovering presence seemed only to heighten his anxiety. His mother was a short, quiet woman of slight build who spoke no English and who dutifully waited on her son for his every need, even when he should have been able to help himself. Juan always acted as if her services were his natural due, and most likely, such suffocating maternal attention was the source of much of his inability to care for himself on even the simplest emotional levels. Through an interpreter, I had on several occasions explained Juan's tenuous prognosis to his mother, and she seemed to understand with a simple fatalism that her son had AIDS and would probably die from it soon.

"I pray to God every day to heal him, to stop his diarrhea, to help him eat so he will get stronger, but he only becomes weaker and weaker..." Here her voice would trail off. "I must learn to accept God's will."

Juan always seemed totally dependent upon his mother in every way imaginable—physically, emotionally, and spiritually. It was as if he had regressed to infancy: he had become completely helpless in the face of disease and impending death.

Juan's two older brothers would accompany their mother to the hospital every so often, but like their mother, they never said much. Rather, they would just quietly stand at a distance in the shadows of the darkened room, somber witnesses to their mother's silent ministrations to their sick brother. When an AIDS patient is slowly dying in the hospital with family members expectantly hovering about, a largely unspoken family psychodrama is often being played out, one of long-standing family conflicts, of unfulfilled hopes, of bitter regrets, and of unspoken love. The tragedy is that by the time such a psychodrama commences, it is often too late to resolve the conflicts and the regrets. And, most sadly, it is often too late to realize the unfulfilled hopes and to give voice to the previously unspoken love. As they maintained their vigil at Juan's bedside, his mother and two brothers had thoughts and feelings that could have filled volumes. The silence in Juan's room was the fearsome silence of his terror, the pathetic silence of his mother's mute and useless ministrations, the sad silence of a family's past that would never be resolved.

On the other side of the wall, in Evaristo's room, there was a great silence of a different sort—the awesome, becalming silence of a dying man at peace with himself and with life. From the outset, Evaristo, too, understood the gravity of his condition, but refused to be afraid or to give up on life. He made it clear that he wanted whatever treatment I felt was appropriate, but he also did not want any pain and unnecessary suffering. Signing a DNR soon after admission, Evaristo had an attitude toward death that was neither flippant nor overly serious.

"I'm ready to die. I'm not afraid. I've done a lot of things in life I'm not proud of, Doc, but I've made my amends and I'm ready when it's my time." His quiet statement was made with neither bravura nor pride.

I am usually able to tell whether a patient is just putting on a brave face to hide inner fear. In Evaristo I sensed no such false courage. Here was a peaceful, yet courageous, acceptance of death that I had rarely come across. Somehow, his matter-of-fact serenity made me feel strangely unsettled: his unaffected calm made me feel as if I were talking to a holy man, not a state prisoner. I envied his equanimity, hoping he might give me insight into my own anxieties about my final journey. What irony, I would often think to myself, that a work-release prisoner, whose life had been so different from mine, would have a peace that would affect me so much. Feeling reassured by his attitude toward death, I always looked forward to seeing Evaristo on rounds, more than anything to marvel at his calm toward dying.

Indeed, more for my own reassurance than anything else, I would still make my routine attempts to offer Evaristo emotional support. Every once in a while, on rounds, I would ask how he was coping with his disease, and perceptively, he would sense my worry and, reversing roles, would end up giving *me* the emotional support.

"Don't worry, Doc," Evaristo would patiently reply with unmistakable resolve, "it's okay—I'm fine with everything . . . it's going to be fine."

Most likely, Evaristo had long ago learned that the only person he could really count on for genuine emotional support was himself. Never one for talking much about his life, he seemed to be a loner. His family was in Puerto Rico, and his only friend in America was a girlfriend, who never visited him in the hospital. He steadfastly refused offers to contact his family or friend—"It would just upset them, and there's no sense in worrying them." Fear, self-pity, regrets about life—these emotions never surfaced in the almost surreal quietude of Evaristo's room.

As I pondered this man's effect on me, I gradually realized that Evaristo's approach to his illness was somewhat paradoxical. He had not given up on life, in that he was willing to allow diagnostic tests such as bronchoscopy and treatments such as chemotherapy for his KS, but he was not fighting death either. As he would put it, "Like most things in life, dying is no big deal." For Evaristo, the fear of death was not going to affect his living. Somehow, he instinctively sensed that the dreaded expectation of death was infinitely greater than the actual event.

Juan, on the other hand, was giving up on life *and* was simultaneously fearful of death, this insoluble conflict causing great misery for both himself and his family. As he became sicker, he refused to allow blood work, to take any of his medicines, or even to eat.

"What's the use—I'm dying, aren't I?" he would whimper on the several occasions I would encourage him to eat and take his medicines.

Yet while giving up on life was not necessarily wrong for Juan—after all, he *was* dying—he continued to recoil from death with blind terror that became more obsessive the sicker he became. Juan had descended into that awful twilight zone between life and death, where every waking moment was consumed by fear and self-pity, and every fitful sleep was haunted by nightmares of death. Juan's being had been reduced to a twisted existential axiom: I fear death, therefore I am.

Juan's final days were disquieting. Never recovering from his liver disease, he remained deeply jaundiced. His scaly, wrinkled skin started to peel in fine flakes, and the yellow hue of his jaundice darkened to a ghostly amber color: he seemed to age decades. Next to fail were his kidneys, the output from his urinary catheter dwindling down to a few drops of tea-colored, highly concentrated urine each nursing shift. Intravenous fluids and diuretics failed to unclog his damaged kidneys, and body toxins rapidly accumulated, dulling his mind and making him increasingly confused. He no longer knew where he was and had trouble recognizing me and his family.

Drifting in and out of sleep, Juan would spend his ever-diminishing hours of semiconsciousness in a groggy stupor, all the time fixing his terrified, glassy-eyed stare onto the darkened ceiling above the bed. Sometimes he would moan—whether in pain, no one could tell for sure. Other times he would call out for his mother—exactly for what, no one could likewise tell with certainty. But most of the time he would spend his dwindling periods of wakefulness looking blankly ahead, oblivious to my questions and to his mother's urgent inquiries into his condition.

Juan's diarrhea soon assumed diluvial proportions, and despite multiple bedding changes every day, he was almost always lying in a pool of liquid brown stool, its odor drifting into the hallway outside his room. The body toxins from kidney and liver failure made his nervous system hyperexcitable, with uncontrollable jerking and twitching of his arms and legs. Predictably, coma and grand mal seizures soon followed, which could be only partially controlled with anticonvulsant medication.

The last organ system to go was Juan's lungs, which became infected by the virulent microorganisms that lurk everywhere in a busy city hospital. Comatose, too malnourished even to mount a reflexive cough, and without any immune system, Juan was incapable of clearing the infected secretions building up in his lungs. Dubbed generations ago "the old man's friend"—

because of the merciful death it gave lingering elderly patients in the pre-antibiotic era—bacterial pneumonia had become in the 1990s the AIDS-related complication that frequently struck down life way too early. As the pneumonia progressed over the ensuing days, Juan's breathing quickened and acquired the gurgling sound of a person drowning in his own lung secretions. The obligatory oxygen and antibiotics I ordered for his pneumonia were a useless formality.

Yet even after Juan had slipped into coma, his face still seemed seized with anguish. His glassy, wide-eyed stare became more terrified and terrifying. It was as if his coma had paradoxically opened up for him a window into a private hell of the unknown, a window to which only he was privy, as his perpetual stare remained locked onto the shadowy, cracked ceiling above his bed. Every once in a while, he would grimace—an involuntary brain-stem reflex, I would have to remind myself—and would become transfixed with a demented smile that seemed creepy and almost demonic, an impression enhanced by occasional low-pitched moans that sounded mournfully desolate. Juan's coma did not appear to provide him refuge from his fears; rather, it seemed more like an unending nightmare of torments.

Throughout Juan's final ordeal, his mother continued her silent bedside vigil, helping the nurses with his care when she was able, but mostly just sitting in a corner, where she would quietly pray or recite the rosary in Spanish. Through an interpreter, I would periodically update her on Juan's condition, but the enormity of his situation was rapidly growing beyond her understanding. All she knew for sure was that her son was dying and that, in her words, only "a great miracle from heaven" could save him. Always appearing emotionless and never having any questions, she did once mention to me how old Juan looked. "He is like an old man," she remarked sadly, with innocent amazement about how AIDS not only took sons before their mothers, but also often made the dying young look so much older than their parents.

I did not dread visiting Juan on my daily rounds, but the incredible fear he had suffered from the onset of his hospitalization saddened me. I sensed that no emotional or spiritual balm could lessen Juan's unbearable psychic pain. Just as I daily took away from Evaristo's room a resolve to learn from his inner strength in the face of death, so I also left Juan each day with a cautionary sense that I must never allow myself to become consumed with fear of death. As Evaristo had instructed me, the dreaded expectation almost always surpasses the actual event, which is far more traumatic for the loved ones than for the one departing. In contemplating Evaristo's reassuring words, I often recalled the old hit song "Is That All There Is?" and mused on how most of us who so fear death would someday, in one form or another, be singing it to ourselves as death embraced us. Indeed, it was my hope that at the moment of release from this life, Juan, too, would discover, however belatedly, Evaristo's insight into the ultimate triviality of death.

When he had been awake and alert earlier during his hospitalization, Juan had steadfastly resisted my attempts to discuss cardiopulmonary resuscitation. Talking about life-support systems, DNR, and "dying in peace" absolutely mortified him, no matter how diplomatic and reassuring I would try to be. As usual, he passively deferred to his mother: "I want my mother to decide about that [resuscitation] when the time comes." Determined to keep Juan out of intensive care, I seized on his insistence that his mother make medical decisions for him and, before coma supervened, got him to sign a health-care proxy that would allow her to decide medical issues for him when he was no longer able to do so. Once Juan slipped into coma, I immediately shoved a health-care proxy DNR form under his mother's nose, and after a perfunctory discussion—"You need to sign this paper so Juan can die in peace"—she readily acquiesced and signed the DNR, even though she probably had little idea what cardiopulmonary resuscitation was. Ethicists and lawyers would have disapproved of my technique, but my controversial approach saw to it that Juan's final days of life would not end on a life-support system. I was determined he would not become a pitiable heart-lung specimen, encased by a disintegrating carcass of skin and bones. And his distraught mother would not have to witness her son being pulverized in the high-tech maws of the intensive care unit, simply to satisfy New York's CPR law.

Juan was found dead in bed at 4:45 A.M. on the twentieth day of his final hospitalization. Last seen "alive" by his night nurse at approximately 1:00 A.M., he was alone at the time of death, which occurred sometime between 1:00 and 4:45 A.M. The on-call intern was paged to examine Juan, officially to pronounce him dead, but as often happens, the intern never showed up. Three and a half hours after Juan's death, as I arrived on the floor to start my day, I was asked by the nurses to "pronounce" Juan, despite the fact he had been lying lifeless in his bed for over three hours. Rather than curse the intern's laziness and grouse about the charade I was asked to perform, I readily agreed, more out of morbid curiosity than anything else.

Juan's room was uncharacteristically sunlit; the nurses had opened the curtains, which had always been closed when Juan was alive. Despite the lack of a doctor's formal death pronouncement, the night nurses had already stuffed his corpse into a hospital body bag, a gray, opaque bag similar to a garment bag. It was odd to see the occupied, zipped-up body bag laid out on the bed, on which were also piled Juan's few personal effects—a few toiletry items, a Sacred Heart of Jesus picture that had been taped on the adjacent wall, a Bible, his threadbare blue quilt, and a hand-embroidered pillow. I was surprised that the nurses, in their efficient zeal, had not already stuffed Juan's things into the usual plastic garbage bag for his family to pick up later. Perhaps, I thought, the night staff felt a tinge of guilt in presumptively enshrouding Juan in a body bag before he had been formally declared dead by a doctor. Although I had pronounced many patients dead, this was the first time I had to open a body bag to do so.

Juan was dead all right, and he looked dreadful. His skin was cold and ashen, and his face was frozen in its familiar stare of sheer terror. With vacant eyes still wide open and rotting teeth fully exposed in a rabid grimace, Juan's preskeletal death mask was both horrific and manifestly sad to behold. It seemed to reflect the sum total of all the fear he had suffered. For some unknown reason—perhaps my final attempt to put the poor man at rest—I tried to close his eyelids, but they would not stay shut.

Next door, Evaristo had no knowledge of Juan's death drama and was in fact in the final days of his own struggle. Whereas Juan's inexorable downhill course left him no hope, Evaristo's condition actually improved after his first dose of chemotherapy. His fever abated, his breathlessness diminished, and his blood-oxygen measurements normalized. The purple KS growth on his eyelid began to regress, and within a week of treatment he was not only able to walk in his room without oxygen but was also regaining his appetite. This gratifying reprieve could have been due to the Bactrim or to the chemotherapy, but I felt that the KS chemotherapy was primarily responsible.

Although never morose, Evaristo's spirits picked up considerably as his breathing improved, and his quiet fatalism yielded to guarded optimism that maybe, just maybe, he might make it out of the hospital alive. He even hoped out loud that he might make it to his parole date coming up in a few weeks. I, too, shared his optimism, but I reminded him that remissions from lung KS often require subsequent treatments and often last only a few months. Evaristo, of course, was typically philosophic about my cautionary caveats.

"Doc, I've learned a long time ago to live one day at a time. If I have a few extra months to live, I can use them, but if I die today, I'm ready. Do your best and keep me comfortable when it's time." I continued to marvel at this man's willingness to accept death as a natural part of living. I really wanted him to make it.

When Evaristo received his dose of chemotherapy for KS, the Spellman hematology consultant, who had prescribed the chemotherapy drugs, advised me that in two to three weeks after this initial treatment, Evaristo should be reevaluated for a possible second dose of chemotherapy. Unlike the treatment of many cancers, such as lymphomas and leukemias, chemotherapy regimens for KS have largely been empiric—that is, seat-of-the-pants—and decisions to treat or retreat a patient's KS depend on the patient's response to initial therapy. By two and a half weeks after the first dose of chemotherapy, Evaristo was beginning to show subtle yet definite signs of relapse: low-grade fevers were being recorded, a mild cough had resurfaced, and his appetite was on the wane again. Although he appeared comfortable at rest, he was walking less often in his room and was spending more and more time in bed. Not having specialized training to prescribe chemotherapy, I needed to reconsult the Spellman hematologist, especially since I felt Evaristo's improvement up to that point had been due to the chemotherapy and not the Bactrim, which, in any case, was about to be discontinued after the standard three weeks' course of anti-PCP treatment.

Spellman's medical consultants, most of whom were part-time and not based at St. Clare's, infrequently revisited patients after the initial consultation, and when subsequent problems arose, it was necessary to recontact the specialist for a follow-up evaluation. When I called the Spellman hematologist for advice about whether further chemotherapy was advisable for Evaristo, I had to leave on his office's answering machine a message to call me as soon as possible. Two days later, Evaristo's condition was slightly worse—he was weaker and now needed oxygen at night—but the hematologist had not yet called me back. On the outside chance that Evaristo's deterioration over the past week was actually due to a flare-up of PCP and not KS, I started him on intravenous pentamidine, a more toxic anti-PCP medication.

"I think we're back to where we started," Evaristo calmly remarked on rounds that day. I agreed, knowing that recurrent lung KS was bad news that needed prompt attention.

Concerned that I had not heard from the hematology consultant, I checked with the Spellman administrative office and confirmed I indeed had the right phone number and that, as far as they knew, the hematologist had not left on vacation or signed out to a colleague. Calling back the hematologist and reaching the same answering machine, I left an urgent message to please call me immediately, also giving my home phone number. At that point, I would have taken the initiative and given Evaristo the same chemotherapy he had received three weeks earlier, but the hospital pharmacy refused to release the drugs unless they received at least a verbal okay from the Spellman hematologist, who, of course, was not returning my repeated calls. The Spellman medical director was en route to Haiti, and the assistant medical director was in the hospital with pneumonia. I was completely on my own. In desperation, I tried to reach the other non-Spellman hematologists at St. Clare's—one said he could not help me, and the other did not return my call.

Before leaving that day, I checked in on Evaristo, more to reassure myself than anything else. Evaristo probably sensed my worry and, as usual, ended up comforting me, rather than the other way around.

"You worry too much, Doc," he scolded me with mock anger. "I'll be okay, and, besides, you've done the best you can." I left the hospital that evening hoping that the hematologist would call me at home and that Evaristo would at least remain stable.

Neither hope materialized. By the next morning, Evaristo was visibly uncomfortable from shortness of breath. Brief conversation or simply rolling over in bed worsened his breathlessness, despite his being on a 100 percent oxygen mask. Small beads of sweat on his forehead were testimony to the incredible work it took him just to breathe in and out. His arms and legs had occasional involuntary jerks from his muscles not getting enough oxygen, and although still alert, he seemed a little confused, his brain struggling to maintain its function in the face of diminishing oxygen supply. Despite his distress, Evaristo's major concern seemed to be comforting and reassuring me that it was all right for him to be dying.

"We gave it a good fight . . . I'm ready, Doc. It's not that big a deal, you know. I'm ready, Doc, and so should you . . ." His voice trailed off to an inaudible whisper and ended with his characteristic smile.

His phrase "and so should you" struck me. On first hearing, it sounded as if he were saying that I should accept *his* impending death, but I suddenly realized that maybe he was reminding me that I should accept *my own* death. I again asked if he wanted anyone notified about his condition, and as before, he declined.

"Got to do this myself . . . don't need any help."

The scant conversation had tired him out, and holding his hand—probably more for my reassurance than his—I again promised to do everything to keep him comfortable. As I squeezed his palm gently, I wished to myself that he had more strength, so that I could learn more about this remarkable man.

To relieve Evaristo's breathlessness, I immediately began a morphine drip, a continuous intravenous infusion of morphine. Later that day I stopped in to see him again. He was almost asleep and appeared more comfortable. His only words on that visit—his final words to me—were, "Thanks, Doc," as I was leaving the room. By then, I wanted Evaristo to die quickly and painlessly.

Evaristo quietly slipped away a few hours later, just two days and six hours after Juan had died. Without ceremony, Evaristo's body was packaged into a body bag, which was put on a stretcher and shipped down to the morgue. As the attendants were wheeling him onto the elevator, I felt no sadness. Rather, I felt privileged to have known him, and I vowed never to forget his courage. I mused on Evaristo's contention that "it's not that big a deal"—that the grim reaper so feared by Juan is nothing but a mirage that evaporates the closer one gets to it.

The next day the Spellman hematologist finally returned my call. Profusely apologetic, he admitted he had "forgotten" to tell anyone at Spellman about being on vacation the prior week and had not arranged coverage during his absence. Unfortunately, such lapses in coverage were not uncommon with Spellman consultants, as any Spellman attending physician could verify. Although angry, I also realized that there was no way to know if Evaristo would have responded to a second course of chemotherapy. Even if he had responded, the second remission would probably have been even briefer than the first. But these were all questions that could never be answered since Evaristo was dead, without the benefit of further chemotherapy.

Later the next day, Abby A. mentioned how the city and New York Department of Corrections were battling over whose responsibility it was to pay for Evaristo's burial. By coincidence too poignant to be contrived, Evaristo had died on his parole date, when he would then no longer be a ward of the state. Each bureaucracy—the city and the state—was trying to "turf" the body off onto the other. Evaristo would have been amused at the fuss.

9

THE LEAST OF THESE MY BRETHREN

When near your death a friend
Asked you what he could do,
"Remember me," you said.
We will remember you.
—*Thom Gunn*

ANOTHER THREE months passes quickly, and photocopied notices are once again posted throughout the hospital: it is time for another Spellman memorial service.

Somehow, it always seems too soon; the last service seems so recent, more like a few weeks, and not three months ago. Perhaps a reflection of the immediacy of my ward work, it is never possible for me, unprompted, to remember even a fraction of my patients who have died over the previous quarter. At best, I could maybe recall those who expired within the prior week or so, but before that, everything is always an amorphous blur. Only the Reading of Names jump-starts my sluggish memory.

From service to service, the names pour forth from the chapel's altar with a rhythmic regularity that seems as predictable as the change of seasons. Indeed, both the names and the seasons sometimes seem to meld together: a patient initially admitted in summertime might linger on until dying the following winter; or, many years, and many memorial services, might elapse between a Spellman patient's first and last admissions, the many intervening sojourns on Spellman blurring together into an indistinct montage of memories. Yet, despite the haziness engendered by the seemingly endless procession of memorial services, the memories of the deceased *do* endure.

"Jose T." . . . the desolate ex-prisoner, whose admission examination was repeatedly delayed that hot August day by the competing demands of my AIDS ward. His pneumonia turned out to be MDR-TB, which never completely responded to multiple medications. His mother hurried from Puerto Rico as soon as she heard from her son, after twelve years' silence. Always at his bedside, she comforted him over the six months he slowly deteriorated, the first love he had experienced in decades.

"Georgie J." . . . Spellman's mercurial Korean-Argentinean partial transsexul. Pascal's "prettiest hooker in Times Square" suffered a bizarre death the evening she went AWOL, a few days after her fourth and final "baptism" at St. Clare's. She impulsively jumped in front of an approaching subway train to retrieve her cosmetics bag, which had been thrown onto the tracks by a drag queen friend during a heated argument over a potential trick. Pascal could not be there for the death of her "prettiest girl, with her big heart."

"Doris J." . . . the single mother with three little girls whose Christmas was rescued by Pascal's gifts. Pascal rescued Doris's "book" immediately after her death from her eighteenth bacterial pneumonia in two years. Her dog-eared brown spiral notebook containing her journal and letters to her three young daughters has been photocopied, and the original and copies entrusted to their child welfare worker, who is also a foster parent for one of the girls.

"Elizabeth Q." . . . the childlike crack addict who pilfered dirty needles from the sharps container, to sell on the streets for cocaine. She did stay out of the sharps container after my bribing her with Xanax, but the following week she never returned from one of her drug runs. A few months later, word reached me that she had recently died at another hospital from pneumonia and endocarditis.

"Luis R." . . . the demented ex-prisoner with a brain abscess and a brother who loved him. He was eventually transferred to Terence Cardinal Cooke Hospice, where he died peacefully two months later. His brother Carlos was at his bedside, rubbing his back with skin lotions, right up to the end. Carlos now does volunteer work on weekends at Spellman.

Equally important as the names that are called out from the chapel's altar are those that are not. The unspoken names of the Spellman living—the many, more numerous patients who are still living with AIDS—are also always present at each memorial service, their lessons speaking forth with the same authority as those of the deceased. In fact, at St. Clare's, little distinction is made between the living and the dead, since—by force of their collective memories—all rapidly meld into one family, as they should. In day-to-day conversations among the staff, recollections of both the living and the dead are often discussed in the present tense, but with a tentativeness that, paradoxically, acknowledges both the fragility of the living and the invincibility of the dead.

The burgeoning memories evoked by the names of the dead at each memorial service bear testimony to the witnesses who remain.

Evelyn T. . . . the work-release prisoner who announced to the TV audience she was "more than a virus." After the TV interview, she was rewarded for her fifteen minutes of fame with a job offer in the mail room of a Midtown law office, which she accepted immediately after parole. Soon thereafter she regained custody of her four-year-old daughter and now lives in housing arranged by the Jewish Board of Family and Children's Services. Evelyn is beginning night classes at CUNY, hoping to become a paralegal secretary.

Israel T. . . . the hopeless, bitter patient who was distraught about his sick son, Samuel, who was in an ICU upstate. He was arrested for parole violation after going AWOL from St. Clare's, in an unsuccessful attempt to see his son in Buffalo. Israel is now back in Sing Sing, refusing all medication and medical care. His son died shortly after he went AWOL, but he does not know about the death.

Rita B. . . . the hotel maid who looks after her mother with Alzheimer's and does her Lord's work in her church. She continues with robust health, always remembering Sister Pascal and her work at St. Clare's in her prayers every evening.

Todd S. . . . the Fire Island showgirl with big, hairy tits. He is delighted the Groshong catheter left his breasts undefiled. The home-administered amphotericin therapy has restored his health almost to normal. His father's much anticipated visit, as often happens, was a nonevent, except for his father's worried concern and ongoing support for Todd's AIDS.

Lois M. . . . the homeless alcoholic woman with frostbite and her first baby. Her healthy baby girl, whom Lois named after herself, is HIV-negative, probably because Lois kept her OB appointments and took AZT prophylaxis prescribed by her doctors. Living at Gift of God and regularly attending Alcoholics Anonymous, she has remained sober, but has become increasingly weaker and wasted. She and her caseworker at Spellman are working on finding an adoptive family for little Lois once AIDS makes it necessary.

The myriad names, spoken and unspoken, seem to reverberate within the narrow walls of the chapel, even when I sit there alone, as I sometimes do at the end of a particularly difficult day. There are many sacred places in New York City—St. John the Divine, Riverside Church, St. Thomas Church—but despite their beauty and solemnity, none equal the spirituality of St. Clare's quiet chapel, where the memories—indeed, perhaps the souls—of these largely unheralded people so intensely rarefy the air therein, consecrating this space in a way few places in the city ever are. And these infinite memories all resonate in perfect unity with words that, although uttered several millennia ago, best seem to embrace, in Pascal's word, the "soul" of St. Clare's Spellman Center:

> When the Son of man shall come in his glory . . . then shall he sit upon his throne of glory: and before him shall be gathered all nations . . . then shall the King say unto them on his right hand, "Come ye blessed of my Father, inherit the kingdom prepared for you . . . for I was hungry, and ye gave me meat: I was thirsty and ye gave me drink: I was a stranger, and ye took me in: naked, and ye clothed me: I was sick and ye visited me: I was in prison, and ye came unto me."
>
> Then shall the righteous answer him, saying, "Lord, when did we see thee hungry, and fed thee, or thirsty, and gave thee drink? When did we see thee a stranger, and took thee in, or naked, and clothed thee? Or when saw we thee sick, or in prison, and came unto thee?"
>
> And the King shall answer and say unto them, "Verily I say unto you: inasmuch as ye have done it unto one of the least of these my brethren, ye have done it unto me." (Matthew 25:40)

These words are the crux of it, helping me to make sense of my work at St. Clare's, and hopefully helping others understand why caring for such people—the unwanted and forgotten—is so necessary. These words of Christ's must speak to everyone—regardless of religious beliefs, if any—about what it means to be human. Indeed, slight paraphrasing of these words adds an even greater human dimension that neither presupposes nor excludes any religious implication: "Inasmuch as you have done it unto one of the least of these your brethren, *you have done it unto yourself.*"

If the life of someone like John R.—the prisoner who sodomized a one-year-old girl—is important and precious beyond measure, then so is mine, as well as the lives of my family and friends, indeed, everyone else's. The value of the lives, and memories, of these so-called "least of these" reaffirms the worth of all human life, regardless of the circumstances.

Indeed, the least of these are really my brethren, not pathetic inferiors upon whom I bestow my pity, a pity that distances and separates. The least of these my brethren teach me that, despite the white coat and professional title—despite my materialistic inclinations, my loving friends and family, my refined sensibilities and overreaching pretensions—I, too, am the least of these. I, too, am living on precious, inconstant time, like all of my patients. And perhaps more so than they, I am even less certain when it will be my time.

Yet, despite these sobering lessons, the least of these also admonish me not to fear death. As I sit in the chapel, either alone or communally at a memorial service, the memories and images of my patients reassure me that, if they can face death, then so can I. I will be as special, but no more unique, when it is my time to confront my imminent mortality, to take to bed and join my patients before me, truly to become, at last, the least of these my brethren.